Python极客项目编程

PYTHON PLAYGROUND

[美] Mahesh Venkitachalam 著　王海鹏 译

人民邮电出版社

北京

图书在版编目（CIP）数据

Python极客项目编程 /（美）温科特卡姆
(Mahesh Venkitachalam) 著；王海鹏译. -- 北京：人民邮电出版社，2017.5（2023.1重印）
ISBN 978-7-115-44976-4

Ⅰ. ①P… Ⅱ. ①温… ②王… Ⅲ. ①软件工具-程序设计 Ⅳ. ①TP311.561

中国版本图书馆CIP数据核字(2017)第043822号

版权声明

Simplified Chinese-language edition copyright © 2017 by Posts and Telecom Press.

Copyright © 2016 by Mahesh Venkitachalam. Title of English-language original: Python Playground，ISBN-13: 978-1-59327-604-1，published by No Starch Press.

All rights reserved.

本书中文简体字版由美国 No Starch 出版社授权人民邮电出版社出版。未经出版者书面许可，对本书任何部分不得以任何方式复制或抄袭。

版权所有，侵权必究。

◆ 著　　[美] Mahesh Venkitachalam
　　译　　王海鹏
　　责任编辑　陈冀康
　　责任印制　焦志炜

◆ 人民邮电出版社出版发行　北京市丰台区成寿寺路11号
　　邮编　100164　电子邮件　315@ptpress.com.cn
　　网址　http://www.ptpress.com.cn
　　大厂回族自治县聚鑫印刷有限责任公司印刷

◆ 开本：800×1000　1/16
　　印张：19.5　　　　　　　2017年5月第1版
　　字数：441千字　　　　　2023年1月河北第42次印刷
　　著作权合同登记号　图字：01-2015-8784号

定价：69.00元
读者服务热线：(010)81055410　印装质量热线：(010)81055316
反盗版热线：(010)81055315
广告经营许可证：京东市监广登字 20170147 号

内容提要

Python 是一种解释型、面向对象、动态数据类型的高级程序设计语言。通过 Python 编程，我们能够解决现实生活中的很多任务。

本书通过 14 个有趣的项目，帮助和鼓励读者探索 Python 编程的世界。全书共 14 章，分别介绍了通过 Python 编程实现的一些有趣项目，包括解析 iTunes 播放列表、模拟人工生命、创建 ASCII 码艺术图、照片拼接、生成三维立体图、创建粒子模拟的烟花喷泉效果、实现立体光线投射算法，以及用 Python 结合 Arduino 和树莓派等硬件的电子项目。本书并不介绍 Python 语言的基础知识，而是通过一系列不简单的项目，展示如何用 Python 来解决各种实际问题，以及如何使用一些流行的 Python 库。

本书适合那些想要通过 Python 编程来进行尝试和探索的读者，适合了解基本的 Python 语法和基本的编程概念的读者进一步学习，对于 Python 程序员有一定的启发和参考价值。

致谢

写一本书就像跑马拉松。有人这么对我说过。但我确实知道，写这本书考验了我的耐力极限，没有亲朋好友和家人为我摇旗呐喊，我不可能完成。

首先，我感谢我的妻子 Hema，她不变的爱、鼓励和耐心，贯穿了完成这项工作的整整两年时间。我感谢我的朋友 Raviprakash Jayaraman，他是我所有不确定的项目的"同谋"，也是本书的技术评审，我们一起吃了许多有趣的午餐、看了不少电影，多次去逛 S.P. Road Zoo。我感谢我的朋友 Seby Kallarakkal，他推动我编写这本书，进行了多次有趣的讨论。我很感谢我的朋友 Santosh Hemachandra 博士在快速傅里叶变换上的有益讨论。我感谢 Karthikeyan Chellappa，帮助我测试 Python 模块的安装，和我一起围绕 Kaikondrahalli 湖跑步。我还要感谢 Matthew Denham（我与他在 Reddit 上交谈），他对万花尺（Spirograph）的数学知识提供了帮助。

我感谢 No Starch 出版社的 Tyler Ortman 和 Bill Pollock，他们接受了我编写本书的意向。还要感谢 Serena Yang 编辑本书的专业工作。我感谢 Nicholas Kramer 对本书的技术评阅。

我感谢我的父母 A.V. Venkitachalam 和 N. Saraswathy，他们为我提供的教育远远超越了他们的财务能力。最后，我感谢所有给我启发的老师，我希望一辈子做一名学生。

前　　言

欢迎阅读本书！在本书中，你会看到 14 个令人兴奋的项目，旨在鼓励你探索 Python 编程的世界。这些项目涉及广泛的主题，如绘制类似万花尺的花纹、生成 ASCII 码艺术图、3D 渲染，以及根据音乐同步投射激光图像。除了本身很有趣之外，这些项目的意图是提供一些起点，让你通过扩展每个项目，来探索你自己的想法。

本书的目标读者

本书的目标读者，是所有想知道如何利用编程来理解和探索想法的人。本书的项目假设你了解基本的 Python 语法和基本的编程概念，并假设你熟悉高中数学知识。我已经尽了最大的努力，详细解释了所有项目中需要的数学知识。

本书不会是你的第一本 Python 书。我不会指导你学习基本知识。但我会通过一系列不简单的项目，向你展示如何用 Python 来解决各种实际问题。在学习这些项目时，你将探索 Python 编程语言的细微差别，并学习如何使用一些流行的 Python 库。但也许更重要的是，你将学习如何将问题分解成几个部分，开发一个算法来解决这个问题，然后从头用 Python 来实现一个解决方案。解决现实世界的问题可能很难，因为它们往往是开放式的，并且需要各个领域的专业知识。但 Python 提供了一些工具，协助解决问题。克服困难，寻找实际问题的解决方案，这是成为专家级程序员的旅途中最重要的环节。

本书的内容

让我们快速浏览一下本书各章的内容。

第一部分：热身运动

第 1 章展示了如何解析 iTunes 播放列表文件，并从中收集有用的信息，如音轨长度和共同的音轨。在第 2 章中，我们使用参数方程及海龟作图法，绘制类似万花尺产生的那些曲线。

第二部分：模拟生命

这部分是用数学模型来模拟现象。在第 3 章中，我们将学习如何实现 Conway 游戏的生命游戏算法，产生动态的模式来创建其他模式，以模拟一种人工生命。第 4 章展示了如何用 Karplus-Strong 算法来创建逼真的弹拨音。然后，在第 5 章中，我们将学习如何实现类鸟群算法，模拟鸟类的聚集行为。

第三部分：图像之乐

这部分介绍使用 Python 读取和操作 2D 图像。第 6 章展示了如何根据图像创建 ASCII 码艺术图。在第 7 章中，我们将进行照片拼接。在第 8 章中，我们将学习如何生成三维立体图，它让人产生 3D 图像的错觉。

第四部分：走进三维

这一部分的项目使用 OpenGL 的 3D 图形库。第 9 章介绍使用 OpenGL 创建简单 3D 图形的基本知识。在第 10 章中，我们将创建粒子模拟的烟花喷泉，它用数学和 OpenGL 着色器来计算和渲染。在第 11 章中，我们将使用 OpenGL 着色器来实现立体光线投射算法，来渲染立体数据，该技术常用于医疗影像，如 MRI 和 CT 扫描。

第五部分：玩转硬件

在最后一部分中，我们将用 Python 来探索 Arduino 微控制器和树莓派。在第 12 章中，我们将利用 Arduino，通过一个简单电路读取并标绘传感器数据。在第 13 章中，我们将利用 Python 和 Arduino 来控制两个旋转镜和激光器，生成响应声音的激光秀。在第 14 章中，我们将使用树莓派打造一个基于网络的气象监测系统。

为何选择 Python

Python 是探索编程的理想语言。作为一种多范式语言，在如何组织程序方面，它提供了极大的灵活性。你可以将 Python 视为脚本语言，简单地执行代码，或将其视为过程语言，把程序组织成一组彼此调用的函数，或将其视为面向对象语言，利用类、继承和模块来建立层次结构。这种灵活性让你可以选择最适合特定项目的编程风格。

如果用更传统的语言来开发，如 C 或 C++，你必须先编译和链接代码，然后才能运行它。使用 Python，你可以编辑后直接运行它（在背后，Python 将你的代码编译成中间字节码，然后由 Python 解释器运行，但这些过程对用户是透明的）。在实践中，用 Python 多次修改并运行代码，要容易很多。

此外，Python 解释器是非常方便的工具，可用于检查代码语法，获得模块的帮助，进行快速计算，甚至测试在开发中的代码。例如，我写 Python 代码时，会打开三个窗口：文本编辑器、命令行和 Python 解释器。我在编辑器中写代码时，会在解释器中导入我的函数或类，边开发边测试。

Python 有一组非常小、简单而强大的数据结构。如果你理解了字符串、列表、元组、字典、列表解析和基本控制结构，如 for 和 while 循环，那么你已经开了个好头。 Python 简洁而有表现力的语法，使得我们很容易只用几行代码，就完成复杂的操作。而一旦熟悉 Python 内置的模块和第三方模块，你将拥有大量的工具，用于解决真正的问题，就像本书中介绍的那样。从 Python 中调用 C/C++代码有标准的方式，反之亦然。因为在 Python 中可以找到库来做几乎所有事情，我们很容易在大型项目中组合使用 Python 和其他语言模块。这就是为什么 Python 被认为是了不起的胶水语言，它可以很容易地组合使用不同的软件组件。本书最后的硬件项目展示了 Python 如何与 Arduino 和 JavaScript 代码协作。真实的软件项目经常使用多种软件技术，Python 非常适合这种分层体系结构。

下面的例子展示了 Python 的易用性。在第 14 章中为树莓派天气监控器开发代码时，我看着温度/湿度传感器的示波器输出，写下这一串数字：

```
0011011100000000000110100000000001010001
```

因为我不能用二进制讲话，所以启动了 Python 解释器并输入：

```
>>> str = '0011011100000000000110100000000001010001'
>>> len(str)
40
>>> [int(str[i:i+8], 2) for i in range(0, 40, 8)]
[55, 0, 26, 0, 81]
```

这行代码将 40 位字符串切分转换成 5 个 8 位的整数，这是我可以理解的。上述数据被解释为 55.0%的湿度，温度为 26.0 摄氏度，校验和是 55 + 26 = 81。

这个例子展示了如何实际使用 Python 解释器作为非常强大的计算器。你不必写一个完整的程序就能快速计算，只要打开解释器，就可以开始。这只是我喜欢 Python 的一个原因，原因还有很多，所以我认为你也会喜欢 Python。

Python 的版本

本书基于 Python 3.3.3，但所有代码都与 Python 2.7.x 和 3.x 兼容。

本书的代码

在本书中，我尽了最大的努力引导你详细研究每个项目的代码，一段接一段地进行。你可以自己输入代码，或按"资源与支持"页所述方式获取书中程序的源代码。

接下来，你会看到一些令人兴奋的项目。我希望你玩这些项目时，享受的乐趣和我创造它们时一样多。而且不要忘记，利用每个项目结束时提供的练习进一步探索。我希望你和本书一起度过许多快乐的编程时光！

资源与支持

本书由异步社区出品，异步社区（www.epubit.com）为您提供相关资源和后续服务。

配套资源

本书提供程序源代码。

您可以扫描右侧二维码添加异步助手为好友获取配套资源。

您也可以在异步社区本书页面中单击 配套资源 ，跳转到下载界面，按提示进行操作即可。注意：为保证购书读者的权益，该操作会给出相关提示，要求输入提取码进行验证。

扫码关注（异步图书）
发送"44976"获取
本书配套资源。

提交勘误

作者和编辑尽最大努力来确保书中内容的准确性，但难免会存在疏漏。欢迎您将发现的问题反馈给我们，帮助我们提升图书的质量。

当您发现错误时，请登录异步社区，按书名搜索，进入本书页面，单击"提交勘误"，输入勘误信息，单击"提交"按钮即可（见下图）。本书的作者和编辑会对您提交的勘误进行审核，确认并接受后，您将获赠异步社区的 100 积分。积分可用于在异步社区兑换优惠券、样书或奖品。

扫码关注本书

扫描下方二维码,您将会在异步社区微信服务号中看到本书信息及相关的服务提示。

与我们联系

我们的联系邮箱是 contact@epubit.com.cn。

如果您对本书有任何疑问或建议,请您发邮件给我们,并请在邮件标题中注明本书书名,以便我们更高效地做出反馈。

如果您有兴趣出版图书、录制教学视频,或者参与图书翻译、技术审校等工作,可以发邮件给我们;有意出版图书的作者也可以到异步社区的在线投稿页面(www.epubit.com/contribute)提交投稿。

如果您所在学校、培训机构或企业,想批量购买本书或异步社区出版的其他图书,也可以发邮件给我们。

如果您在网上发现有针对异步社区出品图书的各种形式的盗版行为,包括对图书全部或部分内容的非授权传播,请您将怀疑有侵权行为的链接发邮件给我们。您的这一举动是对作者权益的保护,也是我们持续为您提供有价值的内容的动力之源。

关于异步社区和异步图书

"异步社区"是人民邮电出版社旗下 IT 专业图书社区,致力于出版精品 IT 技术图书和相关学习产品,为作译者提供优质出版服务。异步社区创办于 2015 年 8 月,提供大量精品 IT 技术图书和电子书,以及高品质技术文章和视频课程。

微信服务号

"异步图书"是由异步社区编辑团队策划出版的精品 IT 专业图书的品牌,依托于人民邮电出版社近 30 年的计算机图书出版积累和专业编辑团队,相关图书在封面上印有异步图书的 LOGO。异步图书的出版领域包括软件开发、大数据、AI、测试、前端、网络技术等。

目　　录

第一部分　热身运动

第 1 章　解析 iTunes 播放列表 ……… 3
- 1.1　iTunes 播放列表文件剖析 ……… 3
- 1.2　所需模块 ……… 5
- 1.3　代码 ……… 5
  - 1.3.1　查找重复 ……… 5
  - 1.3.2　提取重复 ……… 6
  - 1.3.3　查找多个播放列表中共同的音轨 ……… 7
  - 1.3.4　收集统计信息 ……… 8
  - 1.3.5　绘制数据 ……… 8
  - 1.3.6　命令行选项 ……… 9
- 1.4　完整代码 ……… 10
- 1.5　运行程序 ……… 13
- 1.6　小结 ……… 14
- 1.7　实验 ……… 14

第 2 章　万花尺 ……… 15
- 2.1　参数方程 ……… 16
  - 2.1.1　万花尺方程 ……… 17
  - 2.1.2　海龟画图 ……… 19
- 2.2　所需模块 ……… 20
- 2.3　代码 ……… 20
  - 2.3.1　Spiro 构造函数 ……… 20
  - 2.3.2　设置函数 ……… 21
  - 2.3.3　restart()方法 ……… 21
  - 2.3.4　draw()方法 ……… 22
  - 2.3.5　创建动画 ……… 22
  - 2.3.6　SpiroAnimator 类 ……… 23
  - 2.3.7　genRandomParams()方法 ……… 24
  - 2.3.8　重新启动程序 ……… 24
  - 2.3.9　update()方法 ……… 25
  - 2.3.10　显示或隐藏光标 ……… 25
  - 2.3.11　保存曲线 ……… 25
  - 2.3.12　解析命令行参数和初始化 ……… 26
- 2.4　完整代码 ……… 27
- 2.5　运行万花尺动画 ……… 32
- 2.6　小结 ……… 33
- 2.7　实验 ……… 33

第二部分 模拟生命

第 3 章 Conway 生命游戏 ························ 37
- 3.1 工作原理 ························ 38
- 3.2 所需模块 ························ 39
- 3.3 代码 ························ 40
  - 3.3.1 表示网格 ························ 40
  - 3.3.2 初始条件 ························ 41
  - 3.3.3 边界条件 ························ 41
  - 3.3.4 实现规则 ························ 42
  - 3.3.5 向程序发送命令行参数 ········ 43
  - 3.3.6 初始化模拟 ························ 43
- 3.4 完整代码 ························ 44
- 3.5 运行模拟人生的游戏 ························ 46
- 3.6 小结 ························ 47
- 3.7 实验 ························ 47

第 4 章 用 Karplus-Strong 算法产生音乐泛音 ························ 49
- 4.1 工作原理 ························ 51
  - 4.1.1 模拟 ························ 51
  - 4.1.2 创建 WAV 文件 ························ 52
  - 4.1.3 小调五声音阶 ························ 53
- 4.2 所需模块 ························ 54
- 4.3 代码 ························ 54
  - 4.3.1 用 deque 实现环形缓冲区 ······ 54
  - 4.3.2 实现 Karplus-Strong 算法 ······ 55
  - 4.3.3 写 WAV 文件 ························ 56
  - 4.3.4 用 pygame 播放 WAV 文件 ························ 56
  - 4.3.5 main()方法 ························ 57
- 4.4 完整代码 ························ 58
- 4.5 运行拨弦模拟 ························ 61
- 4.6 小结 ························ 62
- 4.7 实验 ························ 62

第 5 章 类鸟群：仿真鸟群 ························ 63
- 5.1 工作原理 ························ 64
- 5.2 所需模块 ························ 64
- 5.3 代码 ························ 64
  - 5.3.1 计算类鸟群的位置和速度 ······ 65
  - 5.3.2 设置边界条件 ························ 66
  - 5.3.3 绘制类鸟群 ························ 67
  - 5.3.4 应用类鸟群规则 ························ 68
  - 5.3.5 添加个体 ························ 70
  - 5.3.6 驱散类鸟群 ························ 71
  - 5.3.7 命令行参数 ························ 71
  - 5.3.8 Boids 类 ························ 71
- 5.4 完整代码 ························ 72
- 5.5 运行类鸟群模拟 ························ 75
- 5.6 小结 ························ 76
- 5.7 实验 ························ 76

第三部分 图片之乐

第 6 章 ASCII 文本图形 ························ 79
- 6.1 工作原理 ························ 80
- 6.2 所需模块 ························ 81
- 6.3 代码 ························ 81
  - 6.3.1 定义灰度等级和网格 ········ 82
  - 6.3.2 计算平均亮度 ························ 82
  - 6.3.3 从图像生成 ASCII 内容 ······ 83
  - 6.3.4 命令行选项 ························ 84
  - 6.3.5 将 ASCII 文本图形字符串写入文本文件 ························ 84

6.4 完整代码 ………………………… 85
6.5 运行ASCII文本图形生成程序 …… 87
6.6 小结 …………………………… 87
6.7 实验 …………………………… 88

第7章 照片马赛克 …………………… 89
7.1 工作原理 ……………………… 90
　　7.1.1 分割目标图像 ……………… 90
　　7.1.2 平均颜色值 ………………… 91
　　7.1.3 匹配图像 …………………… 91
7.2 所需模块 ……………………… 92
7.3 代码 …………………………… 92
　　7.3.1 读入小块图像 ……………… 92
　　7.3.2 计算输入图像的平均
　　　　　颜色值 ……………………… 93
　　7.3.3 将目标图像分割成网格 …… 93
　　7.3.4 寻找小块的最佳匹配 ……… 94
　　7.3.5 创建图像网格 ……………… 95
　　7.3.6 创建照片马赛克 …………… 96
　　7.3.7 添加命令行选项 …………… 97

　　7.3.8 控制照片马赛克的大小 …… 97
7.4 完整代码 ……………………… 98
7.5 运行照片马赛克生成程序 …… 102
7.6 小结 ………………………… 103
7.7 实验 ………………………… 103

第8章 三维立体画 ………………… 105
8.1 工作原理 …………………… 106
　　8.1.1 感知三维立体画中的深度 … 106
　　8.1.2 深度图 …………………… 108
8.2 所需模块 …………………… 109
8.3 代码 ………………………… 109
　　8.3.1 重复给定的平铺图像 …… 109
　　8.3.2 从随机圆创建平铺图像 … 110
　　8.3.3 创建三维立体画 ………… 111
　　8.3.4 命令行选项 ……………… 112
8.4 完整代码 …………………… 113
8.5 运行三维立体画生成程序 … 115
8.6 小结 ………………………… 117
8.7 实验 ………………………… 117

第四部分　走进三维

第9章 理解OpenGL ………………… 121
9.1 老式OpenGL ………………… 122
9.2 现代OpenGL：三维图形管线 … 124
　　9.2.1 几何图元 ………………… 124
　　9.2.2 三维变换 ………………… 125
　　9.2.3 着色器 …………………… 127
　　9.2.4 顶点缓冲区 ……………… 128
　　9.2.5 纹理贴图 ………………… 129
　　9.2.6 显示OpenGL ……………… 129
9.3 所需模块 …………………… 130
9.4 代码 ………………………… 130
　　9.4.1 创建OpenGL窗口 ………… 130
　　9.4.2 设置回调 ………………… 131

　　9.4.3 Scene类 ………………… 133
9.5 完整代码 …………………… 137
9.6 运行OpenGL应用程序 ……… 142
9.7 小结 ………………………… 143
9.8 实验 ………………………… 143

第10章 粒子系统 …………………… 145
10.1 工作原理 …………………… 146
　　10.1.1 为粒子运动建模 ………… 147
　　10.1.2 设置最大范围 …………… 147
　　10.1.3 渲染粒子 ………………… 149
　　10.1.4 利用OpenGL混合来
　　　　　　创建更逼真火花 ………… 149
　　10.1.5 使用公告板 ……………… 150

10.1.6 生成火花动画……151
10.2 所需模块……151
10.3 粒子系统的代码……151
   10.3.1 定义粒子的几何形状……152
   10.3.2 为粒子定义时间延迟数组……153
   10.3.3 设置粒子初始速度……153
   10.3.4 创建顶点着色器……154
   10.3.5 创建片段着色器……156
   10.3.6 渲染……156
   10.3.7 Camera 类……158
10.4 粒子系统完整代码……158
10.5 盒子代码……164
10.6 主程序代码……166
   10.6.1 每步更新这些粒子……167
   10.6.2 键盘处理程序……168
   10.6.3 管理主程序循环……168
10.7 完整主程序代码……169
10.8 运行程序……172
10.9 小结……172
10.10 实验……172

第 11 章 体渲染……173
11.1 工作原理……174
   11.1.1 数据格式……174
   11.1.2 生成光线……175
   11.1.3 显示 OpenGL 窗口……177
11.2 所需模块……178
11.3 项目代码概述……178
11.4 生成三维纹理……178
11.5 完整的三维纹理代码……180
11.6 生成光线……181
   11.6.1 定义颜色立方体的几何形状……182
   11.6.2 创建帧缓冲区对象……184
   11.6.3 渲染立方体的背面……185
   11.6.4 渲染立方体的正面……185
   11.6.5 渲染整个立方体……186
   11.6.6 调整大小处理程序……187
11.7 完整的光线生成代码……187
11.8 体光线投射……192
   11.8.1 顶点着色器……194
   11.8.2 片段着色器……194
11.9 完整的体光线投射代码……196
11.10 二维切片……199
   11.10.1 顶点着色器……201
   11.10.2 片段着色器……202
   11.10.3 针对二维切片的用户界面……202
11.11 完整的二维切片代码……203
11.12 代码整合……206
11.13 完整的主文件代码……207
11.14 运行程序……209
11.15 小结……210
11.16 实验……210

第五部分 玩转硬件

第 12 章 Arduino 简介……215
12.1 Arduino……216
12.2 Arduino 生态系统……217
   12.2.1 语言……218
   12.2.2 IDE……218
   12.2.3 社区……218
   12.2.4 外设……219
12.3 所需模块……219
12.4 搭建感光电路……219
   12.4.1 电路工作原理……219

12.4.2　Arduino 程序·················220
　　　12.4.3　创建实时图表···············221
　12.5　Python 代码···························222
　12.6　完整的 Python 代码··············224
　12.7　运行程序·······························226
　12.8　小结······································227
　12.9　实验······································227

第 13 章　激光音乐秀························229
　13.1　用激光产生图案·····················230
　　　13.1.1　电机控制·······················230
　　　13.1.2　快速傅里叶变换···········232
　13.2　所需模块·······························233
　　　13.2.1　搭建激光秀···················234
　　　13.2.2　连接电机驱动器···········236
　13.3　Arduino 程序·························237
　　　13.3.1　配置 Arduino 数字
　　　　　　输出引脚·······················238
　　　13.3.2　主循环···························238
　　　13.3.3　停止电机·······················240
　13.4　Python 代码···························240
　　　13.4.1　选择音频设备···············241
　　　13.4.2　从输入设备读取数据···241
　　　13.4.3　计算数据流的 FFT·······242
　　　13.4.4　从 FFT 值提取频率
　　　　　　信息·······························243
　　　13.4.5　将频率转换为电机
　　　　　　速度和方向···················243
　　　13.4.6　测试电机设置···············244
　　　13.4.7　命令行选项···················245
　　　13.4.8　手动测试·······················245
　13.5　完整的 Python 代码··············246
　13.6　运行程序·······························249

　13.7　小结······································250
　13.8　实验······································250

第 14 章　基于树莓派的天气监控器·····253
　14.1　硬件······································254
　　　14.1.1　DHT11 温湿度传感器····254
　　　14.1.2　树莓派···························255
　　　14.1.3　设置树莓派···················255
　14.2　安装和配置软件·····················256
　　　14.2.1　操作系统·······················257
　　　14.2.2　初始配置·······················257
　　　14.2.3　Wi-Fi 设置·····················257
　　　14.2.4　设置编程环境···············258
　　　14.2.5　通过 SSH 连接·············259
　　　14.2.6　Web 框架 Bottle···········259
　　　14.2.7　用 flot 绘制···················260
　　　14.2.8　关闭树莓派···················261
　14.3　搭建硬件·······························262
　14.4　代码······································263
　　　14.4.1　处理传感器数据请求···264
　　　14.4.2　绘制数据·······················264
　　　14.4.3　update()方法·················267
　　　14.4.4　用于 LED 的 JavaScript
　　　　　　处理程序·······················267
　　　14.4.5　添加交互性···················268
　14.5　完整代码·······························269
　14.6　运行程序·······························272
　14.7　小结······································273
　14.8　实验······································273

附录 A　软件安装·······························275
附录 B　基础实用电子学···················281
附录 C　树莓派的建议和技巧···········289

第一部分

热身运动

"在初学者的头脑中有很多可能性，在专家的头脑中，可能性很少。"
——铃木俊隆

第 1 章
解析 iTunes 播放列表

我们的 Python 探险始于一个简单的项目，该项目在 iTunes 播放列表文件中查找重复的乐曲音轨，并绘制各种统计数据，如音轨长度和评分。你可以从查看 iTunes 播放列表格式开始，然后学习如何用 Python 提取这些文件的信息。为了绘制这些数据，要用到 matplotlib 库。

在这个项目中，我们将学习以下主题：

- XML 和属性列表（p-list）文件；
- Python 列表和字典；
- 使用 Python 的 set 对象；
- 使用 numpy 数组；
- 直方图和散点图；
- 用 matplotlib 库绘制简单的图；
- 创建和保存数据文件。

1.1 iTunes 播放列表文件剖析

iTunes 资料库中的信息可以导出为播放列表文件（在 iTunes 中选择 File ▶ Library ▶

Export Playlist）。播放列表文件以可扩展标记语言（XML）写成，这是一种基于文本的语言，旨在分层表示基于文本的信息。它包括一些用户定义的标签所构成的树状集合，标签形如<MyTag>，每个标签可以有一些属性和子标签，其中包含附加的信息。

如果在文本编辑器中打开一个播放列表文件，你会看到类似这样的简化版本：

```
<?xml version="1.0" encoding="UTF-8"?>
❶ <!DOCTYPE plist PUBLIC "-//Apple Computer//DTD PLIST 1.0//EN" "http://www
.apple.com/DTDs/PropertyList-1.0.dtd">
❷ <plist version="1.0">
❸ <dict>
❹ <key>Major Version</key><integer>1</integer>
    <key>Minor Version</key><integer>1</integer>
    --snip--
❺   <key>Tracks</key>
    <dict>
        <key>2438</key>
        <dict>
        <key>Track ID</key><integer>2438</integer>
        <key>Name</key><string>Yesterday</string>
        <key>Artist</key><string>The Beatles</string>
        <key>Composer</key><string>Lennon [John], McCartney [Paul]</string>
        <key>Album</key><string>Help!</string>
    </dict>
    --snip--
</dict>
❻   <key>Playlists</key>
    <array>
        <dict>
            <key>Name</key><string>Now</string>
            <key>Playlist ID</key><integer>21348</integer>
            --snip--
            <array>
              <dict>
                <key>Track ID</key><integer>6382</integer>
              </dict>
              --snip--
            </array>
        </dict>
    </array>
</dict>
</plist>
```

属性列表（P-list）文件将对象表示为字典，<dict> 和 <key> 标签与这种方式有关。字典是把键和值关联起来的数据结构，让查找值变得容易。属性列表文件使用字典的字典，其中和键关联的值往往自身又是另一个词典（甚至一个字典列表）。

<xml>标签确定文件为 XML 文件。在这个开始标签之后，文档类型定义（DTD）定义了 XML 文档的结构❶。如你所见，苹果在该标签中的统一资源定位符（URL）中定义了这种结构。

在❷行，文件声明了顶层<plist>标签，其唯一子元素是字典<dict> ❸。该字典包含了各种键，在❹行，包括 Major Version、Minor Version，等等，但我们的兴趣

在❺行的 Tracks 键。注意，该键对应的值也是一个字典，它将整数的音轨 ID 映射到另一个字典，其中包含 Name、Artist 等元素。音乐收藏中的每个音轨都有唯一的音轨 ID 键。

播放列表顺序在❻行由 Playlists 定义，它是顶层字典的一个子节点。

1.2 所需模块

在这个项目中，我们用内置模块 plistlib 来读取播放列表文件。我们还用 matplotlib 库来绘图，用 numpy 的数组来存储数据。

1.3 代码

该项目的目标是找到你的音乐收藏中的重复乐曲，确定播放列表之间共同的音轨，绘制音轨时长的分布图，以及歌曲评分和时长之间的关系图。

随着音乐收藏不断增加，你总会遇到重复的乐曲。为了确定重复的乐曲，查找与 Tracks 键关联的字典中的名称（前面讨论过），找到重复的乐曲，并用音轨长度作为附加准则来检测重复的乐曲，因为名称相同、但长度不同的音轨，可能是不一样的。

要找到两个或多个播放列表之间共同的音轨，你需要将音乐收藏导出为播放列表文件，收集每个播放列表的音轨名称，作为集合进行比较，通过发现集合的交集来找到共同的音轨。

在收集音乐收藏数据的同时，我们将使用强大的 matplotlib（http://matplotlib.org/）绘图软件包来创建一些图，该软件包由已故的 John Hunter 开发。我们可以绘制直方图来显示音轨时长的分布，绘制散点图来比较乐曲评分与长度。

要查看完整的项目代码，请直接跳到 1.4 节。

1.3.1 查找重复

首先可以用 findDuplicates()方法来查找重复的曲目，如下所示：

```
def findDuplicates(fileName):
    print('Finding duplicate tracks in %s...' % fileName)
    # read in a playlist
❶   plist = plistlib.readPlist(fileName)
    # get the tracks from the Tracks dictionary
❷   tracks = plist['Tracks']
    # create a track name dictionary
❸   trackNames = {}
    # iterate through the tracks
❹   for trackId, track in tracks.items():
        try:
```

```
❺              name = track['Name']
               duration = track['Total Time']
               # look for existing entries
❻             if name in trackNames:
                   # if a name and duration match, increment the count
                   # round the track length to the nearest second
❼                 if duration//1000 == trackNames[name][0]//1000:
                       count = trackNames[name][1]
❽                     trackNames[name] = (duration, count+1)
               else:
                   # add dictionary entry as tuple (duration, count)
❾                 trackNames[name] = (duration, 1)
           except:
               # ignore
               pass
```

在❶行，readPlist()方法接受一个 p-list 文件作为输入，并返回顶层字典。在❷行，访问 Tracks 字典，在❸行，创建一个空的字典，用来保存重复的乐曲。在❹行，开始用 items()方法迭代 Tracks 字典，这是 Python 在迭代字典时取得键和值的常用方法。

在❺行，取得字典中每个音轨的名称和时长。用 in 关键字，检查当前乐曲的名称是否已在被构建的字典中❻。如果是这样的，程序检查现有的音轨和新发现的音轨长度是否相同❼，用//操作符，将每个音轨长度除以 1000，由毫秒转换为秒，并四舍五入到最接近的秒，以进行检查（当然，这意味着，只有毫秒差异的两个音轨被认为是相同的）。如果确定这两个音轨长度相等，就取得与 name 关联的值，这是（duration，count）元组，并在❽行增加计数。如果这是程序第一次遇到的音轨名称，就创建一个新条目，count 为 1❾。

将代码的主 for 循环放在 try 语句块中，这是因为一些乐曲音轨可能没有定义乐曲名称。在这种情况下，跳过该音轨，在 except 部分只包含 pass（什么也不做）。

1.3.2 提取重复

利用以下代码，提取重复的音轨：

```
        # store duplicates as (name, count) tuples
❶       dups = []
        for k, v in trackNames.items():
❷           if v[1] > 1:
                dups.append((v[1], k))
        # save duplicates to a file
❸       if len(dups) > 0:
            print("Found %d duplicates. Track names saved to dup.txt" % len(dups))
        else:
            print("No duplicate tracks found!")
❹       f = open("dups.txt", "w")
        for val in dups:
❺           f.write("[%d] %s\n" % (val[0], val[1]))
        f.close()
```

在❶行，创建一个空列表，保存重复乐曲。接下来，迭代遍历 trackNames 字典，如果 count（用 v[1]访问，因为它是元组的第二个元素）大于 1❷，则将元组（name，count）添加到列表中。在❸行，程序打印它找到的信息，然后用 open()方法将信息存入文件❹。在❺行，迭代遍历 dups 列表，写下重复的条目。

1.3.3 查找多个播放列表中共同的音轨

现在，让我们来看看如何找到多个播放列表中共同的乐曲音轨：

```
def findCommonTracks(fileNames):
    # a list of sets of track names
❶   trackNameSets = []
    for fileName in fileNames:
        # create a new set
❷       trackNames = set()
        # read in playlist
❸       plist = plistlib.readPlist(fileName)
        # get the tracks
        tracks = plist['Tracks']
        # iterate through the tracks
        for trackId, track in tracks.items():
            try:
                # add the track name to a set
❹               trackNames.add(track['Name'])
            except:
                # ignore
                pass
    # add to list
❺   trackNameSets.append(trackNames)
    # get the set of common tracks
❻   commonTracks = set.intersection(*trackNameSets)
    # write to file
    if len(commonTracks) > 0:
❼       f = open("common.txt", "w")
        for val in commonTracks:
            s = "%s\n" % val
❽           f.write(s.encode("UTF-8"))
        f.close()
        print("%d common tracks found. "
            "Track names written to common.txt." % len(commonTracks))
    else:
        print("No common tracks!")
```

首先，将播放列表的文件名列表传入 findCommonTracks()，它创建一个空列表❶，保存从每个播放列表创建的一组对象。然后程序迭代遍历列表中的每个文件。对每个文件，创建一个名为 trackNames 的 Python set 对象❷，然后像在 findDuplicates() 中一样，用 plistlib 读入文件❸，取得 Tracks 字典。接下来，迭代遍历该字典中的每个音轨，并添加 trackNames 对象❹。程序读完一个文件中的所有音轨后，将这个集合加入 trackNameSets❺。

在❻行，使用 set.intersection()方法来获得集合之间共同音轨的集合（用 Python*的运算符来展开参数列表）。如果程序发现集合之间的共同音轨，就将音轨名称写入一个文件。在❼行,打开文件,接下来的两行代码完成写入。使用 encode()来格式化输出,确保所有 Unicode 字符都正确处理❽。

1.3.4 收集统计信息

接下来,用 plotStats()方法,针对这些音轨名称收集统计信息:

```
    def plotStats(fileName):
        # read in a playlist
❶      plist = plistlib.readPlist(fileName)
        # get the tracks from the playlist
        tracks = plist['Tracks']
        # create lists of song ratings and track durations
❷      ratings = []
        durations = []
        # iterate through the tracks
        for trackId, track in tracks.items():
            try:
❸              ratings.append(track['Album Rating'])
                durations.append(track['Total Time'])
            except:
                # ignore
                pass

        # ensure that valid data was collected
❹      if ratings == [] or durations == []:
            print("No valid Album Rating/Total Time data in %s." % fileName)
            return
```

这里的目标是收集评分和音轨时长,然后画一些图。在❶行和接下来的代码行中,读取了播放列表文件,并访问 Tracks 字典。接下来,创建两个空列表,保存评分和时长❷(在 iTunes 播放列表中,评分是一个整数,范围是[0, 100])。迭代遍历音轨,在❸行,将评分和时长添加到相应的列表中。最后,在❹行检查完整性,确保从播放列表文件收集了有效数据。

1.3.5 绘制数据

我们已准备好绘制一些数据了。

```
        # scatter plot
❶      x = np.array(durations, np.int32)
        # convert to minutes
❷      x = x/60000.0
❸      y = np.array(ratings, np.int32)
❹      pyplot.subplot(2, 1, 1)
❺      pyplot.plot(x, y, 'o')
❻      pyplot.axis([0, 1.05*np.max(x), -1, 110])
❼      pyplot.xlabel('Track duration')
❽      pyplot.ylabel('Track rating')

        # plot histogram
        pyplot.subplot(2, 1, 2)
❾      pyplot.hist(x, bins=20)
        pyplot.xlabel('Track duration')
```

```
            pyplot.ylabel('Count')

            # show plot
❿           pyplot.show()
```

在❶行，利用 numpy.array()（在代码中作为 np 导入），将音轨时长数据放到 32 位整数数组中。然后在❷行，利用 numpy，将一个操作应用于数组中的每个元素。在这个例子中，将每个以毫秒为单位的时长值除以值 60×1000。在❸行，将乐曲评分保存另一个 numpy 数组 y 中。

用 matplotlib 在同一图像上绘制两张图。在❹行，提供给 subplot()的参数（即，(2, 1, 1)）告诉 matplotlib，该图应该有两行（2）一列（1），且下一个点应在第一行（1）。在❺行，通过调用 plot()创建一个点，并且 o 告诉 matplotlib 用圆圈来表示数据。

在❻行，为 x 轴和 y 轴设置略微大一点儿的范围，以便在图和轴之间留一些空间。在❼和❽行，为 x 轴和 y 轴设置说明文字。

现在用 matplotlib 的方法 hist()，在同一张图中的第二行中，绘制时长直方图❾。bins 参数设置了数据分区的个数，其中每分区用于添加在这个范围内的计数。最后，调用 show()❿，matplotlib 在新窗口中显示出漂亮的图。

1.3.6　命令行选项

现在，我们来看看该程序的 main()方法如何处理命令行参数：

```
    def main():
        # create parser
        descStr = """
        This program analyzes playlist files (.xml) exported from iTunes.
        """
❶       parser = argparse.ArgumentParser(description=descStr)
        # add a mutually exclusive group of arguments
❷       group = parser.add_mutually_exclusive_group()

        # add expected arguments
❸       group.add_argument('--common', nargs='*', dest='plFiles', required=False)
❹       group.add_argument('--stats', dest='plFile', required=False)
❺       group.add_argument('--dup', dest='plFileD', required=False)

        # parse args
❻       args = parser.parse_args()

        if args.plFiles:
            # find common tracks
            findCommonTracks(args.plFiles)
        elif args.plFile:
            # plot stats
            plotStats(args.plFile)
        elif args.plFileD:
            # find duplicate tracks
            findDuplicates(args.plFileD)
        else:
❼           print("These are not the tracks you are looking for.")
```

本书的大多数项目都有命令行参数。不要尝试手工分析它们并搞得一团糟，要将这个日常的任务委派给 Python 的 argparse 模块。在❶行，为此创建了一个 ArgumentParser 对象。该程序可以做三件不同的事情，如发现播放列表之间的共同音轨，绘制统计数据，或发现播放列表中重复的曲目。但是，一个时间程序只能做其中一件事，如果用户决定同时指定两个或多个选项，我们不希望它崩溃。argparse 模块为这个问题提供了一个解决方案，即相互排斥的参数分组。在❷行，用 parser.add_mutually_exclusive_group()方法来创建这样一个分组。

在❸、❹和❺行，指定了前面提到的命令行选项，并输入应该将解析值存入的变量名（args.plFiles、args.plFile 和 args.plFileD），实际解析在❻行完成。参数解析后，就将它们传递给相应的函数，findCommonTracks()、plotStats()和 findDuplicates()，本章前面讨论过这些函数。

要查看参数是否被解析，就测试 args 中相应的变量名。例如，如果用户没有使用--common 选项（该选项找出播放列表之间的共同音轨），解析后 args.plFiles 应该设置为 None。

在❼行，处理用户未输入任何参数的情况。

1.4 完整代码

下面是完整的程序。在 https://github.com/electronut/pp/tree/master/playlist/，你也可以找到本项目的代码和一些测试数据。

```python
import re, argparse
import sys
from matplotlib import pyplot
import plistlib
import numpy as np

def findCommonTracks(fileNames):
    """
    Find common tracks in given playlist files,
    and save them to common.txt.
    """
    # a list of sets of track names
    trackNameSets = []
    for fileName in fileNames:
        # create a new set
        trackNames = set()
        # read in playlist
        plist = plistlib.readPlist(fileName)
        # get the tracks
        tracks = plist['Tracks']
        # iterate through the tracks
        for trackId, track in tracks.items():
            try:
                # add the track name to a set
```

```python
                trackNames.add(track['Name'])
        except:
            # ignore
            pass
    # add to list
    trackNameSets.append(trackNames)
# get the set of common tracks
commonTracks = set.intersection(*trackNameSets)
# write to file
if len(commonTracks) > 0:
    f = open("common.txt", 'w')
    for val in commonTracks:
        s = "%s\n" % val
        f.write(s.encode("UTF-8"))
    f.close()
    print("%d common tracks found. "
          "Track names written to common.txt." % len(commonTracks))
else:
    print("No common tracks!")

def plotStats(fileName):
    """
    Plot some statistics by reading track information from playlist.
    """
    # read in a playlist
    plist = plistlib.readPlist(fileName)
    # get the tracks from the playlist
    tracks = plist['Tracks']
    # create lists of song ratings and track durations
    ratings = []
    durations = []
    # iterate through the tracks
    for trackId, track in tracks.items():
        try:
            ratings.append(track['Album Rating'])
            durations.append(track['Total Time'])
        except:
            # ignore
            pass
    # ensure that valid data was collected
    if ratings == [] or durations == []:
        print("No valid Album Rating/Total Time data in %s." % fileName)
        return

    # scatter plot
    x= np.array(durations, np.int32)
    # convert to minutes
    x = x/60000.0
    y = np.array(ratings, np.int32)
    pyplot.subplot(2, 1, 1)
    pyplot.plot(x, y, 'o')
    pyplot.axis([0, 1.05*np.max(x), -1, 110])
    pyplot.xlabel('Track duration')
    pyplot.ylabel('Track rating')
```

```python
        # plot histogram
        pyplot.subplot(2, 1, 2)
        pyplot.hist(x, bins=20)
        pyplot.xlabel('Track duration')
        pyplot.ylabel('Count')
        # show plot
        pyplot.show()


def findDuplicates(fileName):
    """
    Find duplicate tracks in given playlist.
    """
    print('Finding duplicate tracks in %s...' % fileName)
    # read in playlist
    plist = plistlib.readPlist(fileName)
    # get the tracks from the Tracks dictionary
    tracks = plist['Tracks']
    # create a track name dictionary
    trackNames = {}
    # iterate through tracks
    for trackId, track in tracks.items():
        try:
            name = track['Name']
            duration = track['Total Time']
            # look for existing entries
            if name in trackNames:
                # if a name and duration match, increment the count
                # round the track length to the nearest second
                if duration//1000 == trackNames[name][0]//1000:
                    count = trackNames[name][1]
                    trackNames[name] = (duration, count+1)
            else:
                # add dictionary entry as tuple (duration, count)
                trackNames[name] = (duration, 1)
        except:
            # ignore
            pass
    # store duplicates as (name, count) tuples
    dups = []
    for k, v in trackNames.items():
        if v[1] > 1:
            dups.append((v[1], k))
    # save duplicates to a file
    if len(dups) > 0:
        print("Found %d duplicates. Track names saved to dup.txt" % len(dups))
    else:
        print("No duplicate tracks found!")
    f = open("dups.txt", 'w')
    for val in dups:
        f.write("[%d] %s\n" % (val[0], val[1]))
    f.close()

# gather our code in a main() function
def main():
```

```python
    # create parser
    descStr = """
    This program analyzes playlist files (.xml) exported from iTunes.
    """
    parser = argparse.ArgumentParser(description=descStr)
    # add a mutually exclusive group of arguments
    group = parser.add_mutually_exclusive_group()

    # add expected arguments
    group.add_argument('--common', nargs='*', dest='plFiles', required=False)
    group.add_argument('--stats', dest='plFile', required=False)
    group.add_argument('--dup', dest='plFileD', required=False)

    # parse args
    args = parser.parse_args()

    if args.plFiles:
        # find common tracks
        findCommonTracks(args.plFiles)
    elif args.plFile:
        # plot stats
        plotStats(args.plFile)
    elif args.plFileD:
        # find duplicate tracks
        findDuplicates(args.plFileD)
    else:
        print("These are not the tracks you are looking for.")

# main method
if __name__ == '__main__':
    main()
```

1.5 运行程序

下面是该程序的运行示例:

```
$ python playlist.py --common test-data/maya.xml test-data/rating.xml
```

下面是输出:

```
5 common tracks found. Track names written to common.txt.
$ cat common.txt
God Shuffled His Feet
Rubric
Floe
Stairway To Heaven
Pi's Lullaby
moksha:playlist mahesh$
```

现在,让我们绘制这些音轨的一些统计数据。

```
$ python playlist.py --stats test-data/rating.xml
```

图 1-1 展示了这次运行的输出。

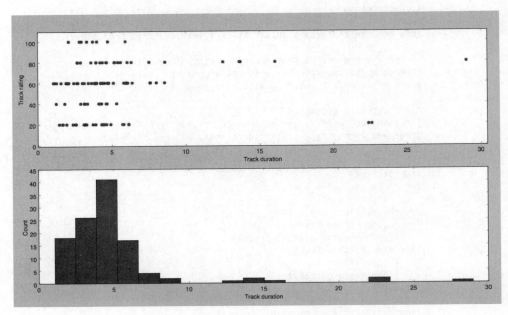

图 1-1 playlist.py 运行示例

1.6 小结

在这个项目中，我们开发了一个程序，分析了 iTunes 播放列表。在这个过程中，我们学习了一些有用的 Python 结构。在接下来的项目中，你将基于这里介绍的一些基础知识，探索各种有趣的主题，深入地研究 Python。

1.7 实验

下面有一些方法可以扩展这个程序。

1. 发现重复音轨时，考虑了以音轨时长作为附加标准，来确定两个音轨是否相同。但寻找共同的音轨时，只用了音轨名称进行比较。在 findCommonTracks()中，请结合音轨时长作为额外的检查。

2. 在 plotStats()方法中，用了 matplotlib 的 hist()方法来计算和显示柱状图。请编写代码手动计算直方图，不用 hist()方法显示。要将结果显示为条形图，请阅读 matplotlib 文档中条形图的部分。

3. 有一些数学公式用于计算相关系数，测量两个变量之间的关系强度。阅读相关性的资料，利用你自己的音乐数据，计算评分/时长散点图中的相关系数。请考虑可以利用播放列表中收集的数据，制作出另外那些散点图。

第2章

万花尺

我们可以用万花尺玩具（如图 2-1 所示）来绘制数学曲线。这种玩具由两个不同尺寸的塑料齿轮组成，一大一小。小的齿轮有几个孔。把钢笔或铅笔放入一个孔，然后在较大齿轮（内部有齿）内旋转里面的小齿轮，保持笔与外轮接触，可以画出无数复杂而奇妙的对称图案。

在这个项目中，我们将用 Python 来创建动画，像万花尺一样绘制曲线。我们的 spiro.py 程序将用 Python 和参数方程来描述程序的万花尺齿轮的运动，并绘制曲线（我称之为螺线）。我们可以将完成的画图保存为 PNG 图像文件，并用命令行选项来指定参数或生成随机螺线。

在这个项目中，我们将学习如何在计算机上绘制螺线。还将学习以下几点：
- 用 turtle 模块创建图形；
- 使用参数方程；
- 利用数学方程来生成曲线；
- 用线段来画曲线；
- 用定时器来生成图形动画；
- 将图形保存为图像文件。

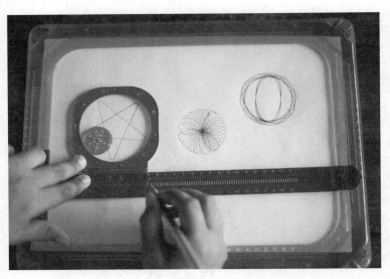

图 2-1　万花尺

关于这个项目要注意：我在这个项目中选择了 turtle 模块用于说明展示，因为它很有趣，但 turtle 比较慢，如果性能很关键，就不适合用它来创建图形（你对海龟有何期望？）。如果想快速画图，有更好的方法，后面的项目将探索一些可选方案。

2.1　参数方程

在本节中，你将看到用参数方程来画圆的简单例子。参数方程将曲线上点的坐标表示为一个变量的函数，该变量称为参数。参数方程让绘制曲线变得容易，因为只要将参数代入方程就能产生曲线。

> **注意**　如果你现在不想学习这部分数学知识，可以跳到下一部分，讨论针对万花尺项目的方程。

我们开始考虑用半径 r 来描述一个圆的方程，圆心位于二维平面的原点。x、y 坐标满足该方程的所有点构成了圆。

现在，请考虑下面的方程：

$$x = r\cos(\theta)$$
$$y = r\sin(\theta)$$

这些方程是圆的参数表示，其中角 θ 是参数。这些方程中（X，Y）的任何值，都满足前面描述的圆的方程，$X^2 + Y^2 = R^2$。如果让 θ 从 0 变到 2π，可以用这些方程来计算圆上对应的 x 和 y 坐标。图 2-2 展示了这种方案。

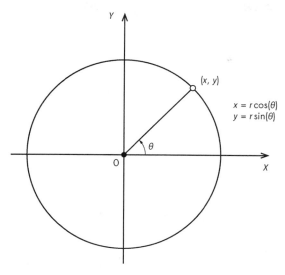

图 2-2 用参数方程描述圆

记住，这两个方程适用于圆心在坐标系原点的圆。将圆心转换到点 (a, b)，就可以将圆置于 xy 平面的任何位置。所以更一般的参数方程就变成 $x = a + r\cos(\theta)$ 和 $y = b + r\cos(\theta)$。现在，让我们来看看描述螺线的方程。

2.1.1 万花尺方程

图 2-3 展示了类似万花尺运动的数学模型。该模型没有齿轮，因为玩具中的齿轮只是为了防止打滑，而在这里不必担心打滑。

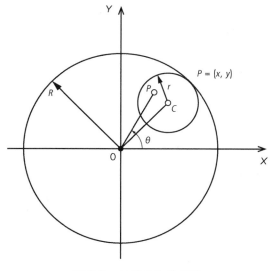

图 2-3 万花尺数学模型

在图 2-3 中，C 是较小的圆的圆心，P 是笔尖。较大的圆半径为 R，较小的圆半径为 r。半径之比表示如下：

$$k = \frac{r}{R}$$

将线段 PC 与小圆半径 r 之比作为变量 l（l = PC / r），它决定了笔尖离小圆圆心有多远。然后，组合这些变量来表示 P 的运动，得到如下的参数方程：

$$x = R\left((1-k)\cos(\theta) + lk\cos\left(\frac{1-k}{k}\theta\right)\right)$$

$$y = R\left((1-k)\sin(\theta) + lk\sin\left(\frac{1-k}{k}\theta\right)\right)$$

> **注意**　这些曲线称为内旋轮线和外旋轮线。虽然方程可能看起来有点吓人，但推导是非常简单的。如果你想探索其中的数学，请参见维基百科。[①]

图 2-4 展示了如何用这些方程，基于参数的变化，产生一条曲线。通过改变参数 R、r 和 l，可以产生变化无穷的迷人曲线。

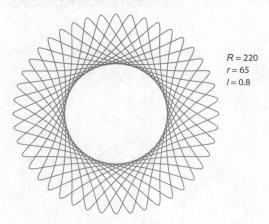

R = 220
r = 65
l = 0.8

图 2-4　示例曲线

将曲线绘制为一系列点之间的线段。如果这些点足够接近，图看起来就像平滑的曲线。真正玩过万花尺就知道，这取决于使用的参数，万花尺可能需要许多转数来完成。要确定何时停止绘图，就要利用万花尺的周期性（即万花尺图案多久开始重复），研究内外圆的半径之比：

---

① http://en.wikipedia.org/wiki/Spirograph/。

$$\frac{r}{R}$$

分子分母除以它们的最大公约数（GCD），化简该分数，分子就告诉我们需要多少圈才能完成曲线。例如，在图 2-4 中，(r, R)的 GCD 是 5。

$$\frac{r}{R} = \frac{65}{220}$$

下面是该分数化简后的形式：

$$\frac{(65/5)}{(220/5)} = \frac{13}{44}$$

这告诉我们，13 圈后，曲线将开始重复。44 告诉我们小圆围绕其中心旋转的圈数，它提示了曲线的形状。在图 2-4 中数一下，会看到图形中花瓣或叶的数目恰好是 44！

一旦用简化形式表示了半径比 r/R，画出螺线的参数 θ 范围就是[0，2πr]。这告诉我们何时停止绘制特定的螺线。不知道该角度的结束范围，就会循环不止，不必要地重复该曲线。

2.1.2 海龟画图

我们可以用 Python 的 turtle 模块来创建图案。这是一个简单的绘图程序，模型是一只海龟拖着尾巴穿过沙滩，留下图案。turtle 模块包括了一些方法，用于设置笔（海龟的尾巴）的位置和颜色，以及其他有用的绘图函数。如你所见，只要少量绘图函数，就可以创建漂亮的螺线。

例如，这个程序用 turtle 画圆。输入以下代码，保存为 drawcircle.py，在 Python 中运行它：

```
  import math
❶ import turtle

  # draw the circle using turtle
  def drawCircleTurtle(x, y, r):
      # move to the start of circle
❷     turtle.up()
❸     turtle.setpos(x + r, y)
❹     turtle.down()

      # draw the circle
❺     for i in range(0, 365, 5):
❻         a = math.radians(i)
❼         turtle.setpos(x + r*math.cos(a), y + r*math.sin(a))

❽ drawCircleTurtle(100, 100, 50)
❾ turtle.mainloop()
```

在❶行，从导入 turtle 模块开始。接下来，定义 drawCircleTurtle()方法，它在❷行调用 up()。这告诉 Python 提笔。换句话说，让笔离开虚拟的纸，这样移动海龟也不会画图。开始绘图之前，先定位海龟。

在❸行，将海龟的位置设置为横轴上的第一个点：(x + r, y)，其中(x，y)是该圆的圆心。现在准备好画图了，所以在❹行调用 down()。在❺行，利用 range(0, 365, 5)开始循环，以 5 为步长递增变量 i，从 0 到 360，变量 i 是角度参数，将传入圆的参数方程，但首先在❻行将它从度转为弧度（大多数计算机程序的角度计算需要弧度）。

在❼行，利用前面讨论过的参数方程计算圆的坐标，并设置相应的海龟位置，这样就从海龟上一个位置画线到新计算的位置（从技术上讲，产生的是 N 边多边形，但因为用了很小的角度，N 将非常大，多边形看起来像一个圆）。

在❽行，调用 drawCircleTurtle()来画圆，在❾行，调用 mainloop()，它保持 tkinter 窗口打开，让你可以欣赏你画的圆（Tkinter 是 Python 默认的 GUI 库）。

现在，我们准备好画一些螺线了！

2.2 所需模块

我们将利用下面的模块创建螺线：
- turtle 模块用于绘图；
- pillow，这是 Python 图像库（PIL）的一个分支，用于保存螺线图像。

2.3 代码

首先，定义类 Sipro，来绘制这些曲线。我们会用这个类一次画一条曲线（利用 draw()方法），并利用一个定时器和 update()方法，产生一组随机螺线的动画。为了绘制 Spiro 对象并产生动画，我们将使用 SpiroAnimator 类。

要查看完整的项目代码，请直接跳到 2.4 节。

2.3.1 Spiro 构造函数

下面是 Spiro 构造函数：

```
# a class that draws a Spirograph
class Spiro:
    # constructor
    def __init__(self, xc, yc, col, R, r, l):
        # create the turtle object
❶       self.t = turtle.Turtle()
        # set the cursor shape
❷       self.t.shape('turtle')
```

```
            # set the step in degrees
❸           self.step = 5
            # set the drawing complete flag
❹           self.drawingComplete = False

            # set the parameters
❺           self.setparams(xc, yc, col, R, r, l)

            # initialize the drawing
❻           self.restart()
```

在❶行,Spiro 构造函数创建一个新的 turtle 对象,这将有助于我们同时绘制多条螺线。在❷行,将光标的形状设置为海龟(在 https://docs.python.org/3.3/library/turtle.html,你可以在 turtle 文档中找到其他选项)。在❸行,将参数绘图角度的增量设置为 5 度,在❹行,设置了一个标志,将在动画中使用它,它会产生一组螺线。

在❺和❻行,调用设置函数,接下来讨论该函数。

2.3.2 设置函数

现在让我们看看 getParams()方法,它帮助初始化 Spiro 对象,如下所示:

```
        # set the parameters
        def setparams(self, xc, yc, col, R, r, l):
            # the Spirograph parameters
❶           self.xc = xc
            self.yc = yc
❷           self.R = int(R)
            self.r = int(r)
            self.l = l
            self.col = col
            # reduce r/R to its smallest form by dividing with the GCD
❸           gcdVal = gcd(self.r, self.R)
❹           self.nRot = self.r//gcdVal
            # get ratio of radii
            self.k = r/float(R)
            # set the color
            self.t.color(*col)
            # store the current angle
❺           self.a = 0
```

在❶行,保存曲线中心的坐标。然后在❷行,将每个圆的半径(R 和 r)转换为整数并保存这些值。在❸行,用 Python 模块 fractions 内置的 gcd()方法来计算半径的 GCD。我们将用这些信息来确定曲线的周期性,在❹行将它保存为 self.nRot。最后,在❺行,保存当前的角度,我们将用它来创建动画。

2.3.3 restart()方法

接下来,restart()方法重置 Spiro 对象的绘制参数,让它准备好重画:

```
        # restart the drawing
        def restart(self):
            # set the flag
```

```
❶          self.drawingComplete = False
            # show the turtle
❷          self.t.showturtle()
            # go to the first point
❸          self.t.up()
❹          R, k, l = self.R, self.k, self.l
            a = 0.0
❺          x = R*((1-k)*math.cos(a) + l*k*math.cos((1-k)*a/k))
            y = R*((1-k)*math.sin(a) - l*k*math.sin((1-k)*a/k))
❻          self.t.setpos(self.xc + x, self.yc + y)
❼          self.t.down()
```

这里用了布尔标志 drawingComplete，来确定绘图是否已经完成，在❶行初始化该标志。绘制多个 Spiro 对象时，这个标志是有用的，因为它可以追踪某个特定的螺线是否完成。在❷行，显示海龟光标，以防它被隐藏。在❸行提起笔，这样就可以在❻行移动到第一个位置而不画线。在❹行，使用了一些局部变量，以保持代码紧凑。然后，在❺行，计算角度 a 设为 0 时的 x 和 y 坐标，以获得曲线的起点。最后，在❼行，我们已完成，并落笔。Setpos()调用将绘制实际的线。

2.3.4　draw()方法

draw()方法用连续的线段绘制该曲线。

```
            # draw the whole thing
            def draw(self):
                # draw the rest of the points
                R, k, l = self.R, self.k, self.l
❶              for i in range(0, 360*self.nRot + 1, self.step):
                    a = math.radians(i)
❷                  x = R*((1-k)*math.cos(a) + l*k*math.cos((1-k)*a/k))
                    y = R*((1-k)*math.sin(a) - l*k*math.sin((1-k)*a/k))
                    self.t.setpos(self.xc + x, self.yc + y)
                # drawing is now done so hide the turtle cursor
❸              self.t.hideturtle()
```

在❶行，迭代遍历参数 i 的完整范围，它以度表示，是 360 乘以 nRot。在❷行，计算参数 i 的每个值对应的 X 和 Y 坐标。在❸行，隐藏光标，因为我们已完成绘制。

2.3.5　创建动画

update()方法展示了一段一段绘制曲线来创建动画时所使用的绘图方法。

```
            # update by one step
            def update(self):
                # skip the rest of the steps if done
❶              if self.drawingComplete:
                    return
                # increment the angle
❷              self.a += self.step
                # draw a step
                R, k, l = self.R, self.k, self.l
                # set the angle
❸              a = math.radians(self.a)
```

```
            x= self.R*((1-k)*math.cos(a) + l*k*math.cos((1-k)*a/k))
            y = self.R*((1-k)*math.sin(a) - l*k*math.sin((1-k)*a/k))
            self.t.setpos(self.xc + x, self.yc + y)
            # if drawing is complete, set the flag
❹           if self.a >= 360*self.nRot:
                self.drawingComplete = True
                # drawing is now done so hide the turtle cursor
                self.t.hideturtle()
```

在❶行，update()方法检查 drawingComplete 标志是否设置。如果没有设置，则继续执行代码其余的部分。在❷行，update()增加当前的角度。从❸行开始，它计算当前角度对应的（X，Y）位置并将海龟移到那里，在这个过程中画出线段。

讨论万花尺方程时，我提到了曲线的周期性。在一定的角度后，万花尺的图案开始重复。在❹行，检查角度是否达这条特定曲线计算的完整范围。如果是这样，就设置 drawingComplete 标志，因为绘图完成了。最后，隐藏海龟光标，你可以看到自己美丽的创作。

2.3.6　SpiroAnimator 类

SpiroAnimator 类让我们同时绘制随机的螺线。该类使用一个计时器，每次绘制曲线的一段。这种技术定期更新图像，并允许程序处理事件，如按键、鼠标点击，等等。但是，这种计时器技术需要对绘制代码进行一些调整。

```
# a class for animating Spirographs
class SpiroAnimator:
    # constructor
    def __init__(self, N):
        # set the timer value in milliseconds
❶       self.deltaT = 10
        # get the window dimensions
❷       self.width = turtle.window_width()
        self.height = turtle.window_height()
        # create the Spiro objects
❸       self.spiros = []
        for i in range(N):
            # generate random parameters
❹           rparams = self.genRandomParams()
            # set the spiro parameters
❺           spiro = Spiro(*rparams)
            self.spiros.append(spiro)
            # call timer
❻       turtle.ontimer(self.update, self.deltaT)
```

在❶行，该 SpiroAnimator 构造函数将 DeltaT 设置为 10，这是以毫秒为单位的时间间隔，将用于定时器。在❷行，保存海龟窗口的尺寸。然后在❸行创建一个空数组，其中将填入一些 Spiro 对象。这些封装的万花尺绘制，然后循环 N 次（N 传入给构造函数 SpiroAnimator），在❺行创建一个新的 Spiro 对象，并将它添加到 Spiro 对象的列表中。这里的 rparams 是一个元组，需要传入到 Spiro 构造函数。但是，构造函数需要一个参数列表，所以用 Python 的*运算符将元组转换为参数列表。

最后，在❻行，设置 turtle.ontimer()方法每隔 DeltaT 毫秒调用 update()。

请注意，在❹行调用了一个辅助方法，名为 genRandomParams()。接下来就看看这个方法。

2.3.7　genRandomParams()方法

我们用 genRandomParams()方法来生成随机参数，在每个 Spiro 对象创建时发送给它，来生成各种曲线。

```
# generate random parameters
  def genRandomParams(self):
      width, height = self.width, self.height
❶     R = random.randint(50, min(width, height)//2)
❷     r = random.randint(10, 9*R//10)
❸     l = random.uniform(0.1, 0.9)
❹     xc = random.randint(-width//2, width//2)
❺     yc = random.randint(-height//2, height//2)
❻     col = (random.random(),
             random.random(),
             random.random())
❼     return (xc, yc, col, R, r, l)
```

为了生成随机数，利用来自 Python 的 random 模块的两个方法：randint()，它返回指定范围内的随机整数，以及 uniform()，它对浮点数做同样的事。在❶行，将 R 设置为 50 至窗口短边一半长度的随机整数，在❷行，将 r 设置为 R 的 10%至 90%之间。

然后，在❸行，将 l 设置为 0.1 至 0.9 之间的随机小数。在❹和❺行，在屏幕边界内随机选择 x 和 y 坐标，选择屏幕上的一个随机点作为螺线的中心。在❻行随机设置为红、绿和蓝颜色的成分，为曲线指定随机的颜色。最后，在❼行，所有计算的参数作为一个元组返回。

2.3.8　重新启动程序

我们将用另一个 restart()方法来重新启动程序。

```
# restart spiro drawing
  def restart(self):
      for spiro in self.spiros:
          # clear
          spiro.clear()
          # generate random parameters
          rparams = self.genRandomParams()
          # set the spiro parameters
          spiro.setparams(*rparams)
          # restart drawing
          spiro.restart()
```

它遍历所有的 Spiro 对象，清除以前绘制的每条螺线，分配新的螺线参数，然后重新启动程序。

2.3.9 update()方法

下面的代码展示了 SproAnimator 中的 update()方法，它由定时器调用，以动画的形式更新所有的 Spiro 对象：

```
    def update(self):
        # update all spiros
❶       nComplete = 0
        for spiro in self.spiros:
            # update
❷           spiro.update()
            # count completed spiros
❸           if spiro.drawingComplete:
                nComplete += 1
        # restart if all spiros are complete
❹       if nComplete == len(self.spiros):
            self.restart()
        # call the timer
❺       turtle.ontimer(self.update, self.deltaT)
```

update()方法使用一个计数器 nComplete 来记录已画的 Spiro 对象的数目。在❶行初始化后，它遍历 Spiro 对象的列表，在❷行更新它们，如果一个 Spiro 完成，就在❸行将计数器加 1。

在循环外的❹行，检查计数器，看看是否所有对象都已画完。如果已画完，调用 restart()方法重新开始新的螺线动画。在❺行 restart()的末尾，调用计时器方法，它在 DeltaT 毫秒后再次调用 update()。

2.3.10 显示或隐藏光标

最后，使用下面的方法来打开或关闭海龟光标。这可以让绘图更快。

```
    # toggle turtle cursor on and off
    def toggleTurtles(self):
        for spiro in self.spiros:
            if spiro.t.isvisible():
                spiro.t.hideturtle()
            else:
                spiro.t.showturtle()
```

2.3.11 保存曲线

使用 saveDrawing()方法，将绘制保存为 PNG 图像文件。

```
    # save drawings as PNG files
    def saveDrawing():
        # hide the turtle cursor
❶       turtle.hideturtle()
        # generate unique filenames
❷       dateStr = (datetime.now()).strftime("%d%b%Y-%H%M%S")
        fileName = 'spiro-' + dateStr
        print('saving drawing to %s.eps/png' % fileName)
```

```
        # get the tkinter canvas
❶       canvas = turtle.getcanvas()
        # save the drawing as a postscipt image
❷       canvas.postscript(file = fileName + '.eps')
        # use the Pillow module to convert the postscript image file to PNG
❸       img = Image.open(fileName + '.eps')
❹       img.save(fileName + '.png', 'png')
        # show the turtle cursor
❺       turtle.showturtle()
```

在❶行，隐藏海龟光标，这样就不会在最后的图形中看到它。然后，在❷行，使用 datetime()，利用当前时间和日期（以"日—月—年—时—分—秒"的格式），以生成图像文件的唯一名称。将这个字符串加在 spiro-后面，生成文件名。

turtle 程序采用 tkinter 创建的用户界面（UI）窗口，在❸和❹行，利用 tkinter 的 canvas 对象，将窗口保存为嵌入式 PostScript（EPS）文件格式。由于 EPS 是矢量格式，你可以用高分辨率打印它，但 PNG 用途更广，所以在❺行用 Pillow 打开 EPS 文件，并在❻行将它保存为 PNG 文件。最后，在❼行，取消隐藏海龟光标。

2.3.12 解析命令行参数和初始化

像第 1 章中一样，在 main()方法中用 argparse 来解析传入程序的命令行选项。

```
❶       parser = argparse.ArgumentParser(description=descStr)

        # add expected arguments
❷       parser.add_argument('--sparams', nargs=3, dest='sparams', required=False,
                            help="The three arguments in sparams: R, r, l.")

        # parse args
❸       args = parser.parse_args()
```

在❶行，创建参数解析器对象，在❷行，向解析器添加--sparams 可选参数。在❸行，调用函数进行实际的解析。

接下来，代码设置了一些 turtle 参数。

```
        # set the width of the drawing window to 80 percent of the screen width
❶       turtle.setup(width=0.8)

        # set the cursor shape to turtle
❷       turtle.shape('turtle')

        # set the title to Spirographs!
❸       turtle.title("Spirographs!")
        # add the key handler to save our drawings
❹       turtle.onkey(saveDrawing, "s")
        # start listening
❺       turtle.listen()

        # hide the main turtle cursor
❻       turtle.hideturtle()
```

在❶行，用 setup()将绘图窗口的宽度设置为 80％的屏幕宽度（你也可以给 setup 指定高度和原点参数）。在❷行，设置光标形状为海龟，在❸行，设置程序窗口的标题为 Spirographs!，在❹行，利用 onkey()和 saveDrawing，在按下 S 时保存图画。然后，在❺行，调用 listen()让窗口监听用户事件。最后，在❻行，隐藏海龟光标。

命令行参数解析后，代码的其余部分进行如下：

```
    # check for any arguments sent to --sparams and draw the Spirograph
❶   if args.sparams:
❷       params = [float(x) for x in args.sparams]
        # draw the Spirograph with the given parameters
        col = (0.0, 0.0, 0.0)
❸       spiro = Spiro(0, 0, col, *params)
❹       spiro.draw()
    else:
        # create the animator object
❺       spiroAnim = SpiroAnimator(4)
        # add a key handler to toggle the turtle cursor
❻       turtle.onkey(spiroAnim.toggleTurtles, "t")
        # add a key handler to restart the animation
❼       turtle.onkey(spiroAnim.restart, "space")
    # start the turtle main loop
❽   turtle.mainloop()
```

在❶行，首先检查是否有参数赋给--sparams。如果有，就从字符串中提取它们，用"列表解析"将它们转换成浮点数❷（列表解析是一种 Python 结构，让你以紧凑而强大的方式创建一个列表，例如，a = [2*x for x in range(1, 5)]创建前 4 个偶数的列表）。

在❸行，利用任何提取的参数来构造 Spiro 对象（利用 Python 的*运算符，它将列表转换为参数）。然后，在❹行，调用 draw()，绘制螺线。

现在，如果命令行上没有指定参数，就进入随机模式。在❺行，创建一个 SpiroAnimator 对象，向它传入参数 4，告诉它创建 4 幅图画。在❻行，利用 onkey()来捕捉按键 T，这样就可以用它来切换海龟光标（toggleTurtles），在❼行，处理空格键（space），这样就可以用它在任何时候重新启动动画。最后，在❽行，调用 mainloop()告诉 tkinter 窗口保持打开，监听事件。

2.4 完整代码

下面是完整的万花尺程序。也可以从 https://github.com/electronut/pp/blob/master/spirograph/spiro.py 下载该项目的代码。

```
import sys, random, argparse
import numpy as np
import math
import turtle
import random
```

```python
from PIL import Image
from datetime import datetime
from fractions import gcd

# a class that draws a Spirograph
class Spiro:
    # constructor
    def __init__(self, xc, yc, col, R, r, l):

        # create the turtle object
        self.t = turtle.Turtle()
        # set the cursor shape
        self.t.shape('turtle')

        # set the step in degrees
        self.step = 5
        # set the drawing complete flag
        self.drawingComplete = False

        # set the parameters
        self.setparams(xc, yc, col, R, r, l)

        # initialize the drawing
        self.restart()

    # set the parameters
    def setparams(self, xc, yc, col, R, r, l):
        # the Spirograph parameters
        self.xc = xc
        self.yc = yc
        self.R = int(R)
        self.r = int(r)
        self.l = l
        self.col = col
        # reduce r/R to its smallest form by dividing with the GCD
        gcdVal = gcd(self.r, self.R)
        self.nRot = self.r//gcdVal
        # get ratio of radii
        self.k = r/float(R)
        # set the color
        self.t.color(*col)
        # store the current angle
        self.a = 0

    # restart the drawing
    def restart(self):
        # set the flag
        self.drawingComplete = False
        # show the turtle
        self.t.showturtle()
        # go to the first point
        self.t.up()
        R, k, l = self.R, self.k, self.l
        a = 0.0
        x = R*((1-k)*math.cos(a) + l*k*math.cos((1-k)*a/k))
        y = R*((1-k)*math.sin(a) - l*k*math.sin((1-k)*a/k))
```

```python
            self.t.setpos(self.xc + x, self.yc + y)
            self.t.down()

    # draw the whole thing
    def draw(self):
        # draw the rest of the points
        R, k, l = self.R, self.k, self.l
        for i in range(0, 360*self.nRot + 1, self.step):
            a = math.radians(i)
            x = R*((1-k)*math.cos(a) + l*k*math.cos((1-k)*a/k))
            y = R*((1-k)*math.sin(a) - l*k*math.sin((1-k)*a/k))
            self.t.setpos(self.xc + x, self.yc + y)
        # drawing is now done so hide the turtle cursor
        self.t.hideturtle()

    # update by one step
    def update(self):
        # skip the rest of the steps if done
        if self.drawingComplete:
            return
        # increment the angle
        self.a += self.step
        # draw a step
        R, k, l = self.R, self.k, self.l
        # set the angle
        a = math.radians(self.a)
        x = self.R*((1-k)*math.cos(a) + l*k*math.cos((1-k)*a/k))
        y = self.R*((1-k)*math.sin(a) - l*k*math.sin((1-k)*a/k))
        self.t.setpos(self.xc + x, self.yc + y)
        # if drawing is complete, set the flag
        if self.a >= 360*self.nRot:
            self.drawingComplete = True
            # drawing is now done so hide the turtle cursor
            self.t.hideturtle()

    # clear everything
    def clear(self):
        self.t.clear()

# a class for animating Spirographs
class SpiroAnimator:
    # constructor
    def __init__(self, N):
        # set the timer value in milliseconds
        self.deltaT = 10
        # get the window dimensions
        self.width = turtle.window_width()
        self.height = turtle.window_height()
        # create the Spiro objects
        self.spiros = []
        for i in range(N):
            # generate random parameters
            rparams = self.genRandomParams()
            # set the spiro parameters
            spiro = Spiro(*rparams)
            self.spiros.append(spiro)
```

```python
            # call timer
            turtle.ontimer(self.update, self.deltaT)

    # restart spiro drawing
    def restart(self):
        for spiro in self.spiros:
            # clear
            spiro.clear()
            # generate random parameters
            rparams = self.genRandomParams()
            # set the spiro parameters
            spiro.setparams(*rparams)
            # restart drawing
            spiro.restart()

    # generate random parameters
    def genRandomParams(self):
        width, height = self.width, self.height
        R = random.randint(50, min(width, height)//2)
        r = random.randint(10, 9*R//10)
        l = random.uniform(0.1, 0.9)
        xc = random.randint(-width//2, width//2)
        yc = random.randint(-height//2, height//2)
        col = (random.random(),
               random.random(),
               random.random())
        return (xc, yc, col, R, r, l)

    def update(self):
        # update all spiros
        nComplete = 0
        for spiro in self.spiros:
            # update
            spiro.update()
            # count completed spiros
            if spiro.drawingComplete:
                nComplete += 1
        # restart if all spiros are complete
        if nComplete == len(self.spiros):
            self.restart()
        # call the timer
        turtle.ontimer(self.update, self.deltaT)

    # toggle turtle cursor on and off
    def toggleTurtles(self):
        for spiro in self.spiros:
            if spiro.t.isvisible():
                spiro.t.hideturtle()
            else:
                spiro.t.showturtle()

# save drawings as PNG files
def saveDrawing():
    # hide the turtle cursor
    turtle.hideturtle()
    # generate unique filenames
```

```python
        dateStr = (datetime.now()).strftime("%d%b%Y-%H%M%S")
        fileName = 'spiro-' + dateStr
        print('saving drawing to %s.eps/png' % fileName)
        # get the tkinter canvas
        canvas = turtle.getcanvas()
        # save the drawing as a postscipt image
        canvas.postscript(file = fileName + '.eps')
        # use the Pillow module to convert the poscript image file to PNG
        img = Image.open(fileName + '.eps')
        img.save(fileName + '.png', 'png')
        # show the turtle cursor
        turtle.showturtle()

# main() function
def main():
    # use sys.argv if needed
    print('generating spirograph...')
    # create parser
    descStr = """This program draws Spirographs using the Turtle module.
    When run with no arguments, this program draws random Spirographs.

    Terminology:

    R: radius of outer circle
    r: radius of inner circle
    l: ratio of hole distance to r
    """

    parser = argparse.ArgumentParser(description=descStr)

    # add expected arguments
    parser.add_argument('--sparams', nargs=3, dest='sparams', required=False,
                        help="The three arguments in sparams: R, r, l.")

    # parse args
    args = parser.parse_args()

    # set the width of the drawing window to 80 percent of the screen width
    turtle.setup(width=0.8)

    # set the cursor shape to turtle
    turtle.shape('turtle')

    # set the title to Spirographs!
    turtle.title("Spirographs!")
    # add the key handler to save our drawings
    turtle.onkey(saveDrawing, "s")
    # start listening
    turtle.listen()

    # hide the main turtle cursor
    turtle.hideturtle()

    # check for any arguments sent to --sparams and draw the Spirograph
    if args.sparams:
        params = [float(x) for x in args.sparams]
```

```
        # draw the Spirograph with the given parameters
        col = (0.0, 0.0, 0.0)
        spiro = Spiro(0, 0, col, *params)
        spiro.draw()
    else:
        # create the animator object
        spiroAnim = SpiroAnimator(4)
        # add a key handler to toggle the turtle cursor
        turtle.onkey(spiroAnim.toggleTurtles, "t")
        # add a key handler to restart the animation
        turtle.onkey(spiroAnim.restart, "space")

    # start the turtle main loop
    turtle.mainloop()

# call main
if __name__ == '__main__':
    main()
```

2.5 运行万花尺动画

现在该运行程序了。

```
$ python spiro.py
```

默认情况下，spiro.py 程序绘制随机螺线，如图 2-5 所示。按 S 键保存绘制。

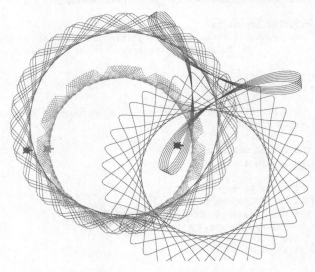

图 2-5　spiro.py 的运行示例

现在，再次运行程序，这次在命令行传入参数，画出特定的螺线。

```
$ python spiro.py --sparams 300 100 0.9
```

图 2-6 展示了输出结果。如你所见，这段代码根据用户指定的参数绘制了一条螺线，图 2-5 和它不同，展示了几个随机螺线的动画。

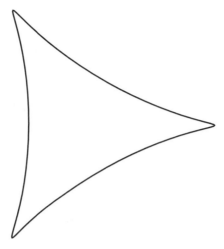

图 2-6　用具体参数运行 spiro.py 的示例

2.6　小结

在这个项目中，我们学习了如何创建万花尺那样的曲线。我们还学习了如何调整输入参数，来生成各种不同的曲线，并在屏幕上产生动画。我希望你喜欢创造这些螺线（在第 13 章你会惊喜地发现，可以学到如何将螺线投影到墙上）。

2.7　实验

下面有一些方法可以进一步尝试螺线。

1. 现在你已知道如何画圆，请写一个程序来绘制随机的对数螺线。找到参数形式的对数螺线方程，然后用它来绘制螺线。

2. 你可能已经注意到，画曲线时，海龟光标总是朝右，但这不是海龟移动的方式！请调整海龟的方向，在绘制曲线时，让它朝向绘制的方向（提示：每步计算连续点之间的方向矢量，用 turtle.setheading()方法来调整海龟的方向）。

3. 尝试用海龟绘制 Koch snowflake（科赫雪花），它是利用递归（即调用自身的函数）的分形曲线。可以像这样组织递归函数调用：

```
# recursive Koch snowflake
def kochSF(x1, y1, x2, y2, t):
    # compute intermediate points p2, p3
    if segment_length > 10:
        # recursively generate child segments
```

```
        # flake #1
        kochSF(x1, y1, p1[0], p1[1], t)
        # flake #2
        kochSF(p1[0], p1[1], p2[0], p2[1], t)
        # flake #3
        kochSF(p2[0], p2[1], p3[0], p3[1], t)
        # flake #4
        kochSF(p3[0], p3[1], x2, y2, t)
    else:
        # draw
        # ...
```

如果你确实遇到困难，可以在 http://electronut.in/koch-snowflake-and-the-thue-morse-sequence/ 找到我的解决方案。

第二部分

模拟生命

"首先,我们假设奶牛是一个球体。"
——佚名物理学笑话

第3章
Conway 生命游戏

你可以用计算机来研究一个系统,方法是建立该系统的数学模型,编程表示该模型,然后让该模型随着时间的推移而演进。有很多种计算机模拟,但我会专注于一个著名的模拟,即 Conway 生命游戏,它是英国数学家 John Conway 的成果。生命游戏是细胞自动机,即网格上的一组彩色细胞,根据定义相邻细胞状态的一组规则,经过一些时间逐步演进。

这个项目将创建一个 N×N 的细胞网格,通过应用 Conway 生命游戏的规则,模拟系统随时间的演进。你将显示每个时间段的游戏状态,并提供一些方式将输出保存到文件。你会设置系统的初始状态,要么是随机分布,要么是预先设计的图案。

该模拟由以下几部分组成:
- 在一维或两维空间中定义的属性;
- 在模拟中的每一步,改变这种属性的数学规则;
- 随着系统的演进,显示或记录系统状态的方式。

在 Conway 生命游戏中的细胞可以处于 ON 或 OFF 状态。游戏从一个初始状态开始,其中每个细胞分配一个状态,数学规则决定其状态如何随时间而改变。Conway 生命游戏中令人惊奇的是,只有 4 个简单的规则,系统演进会产生行为极其复杂的图案,仿佛它们是活的。图案包括"滑翔机",即在网格上滑动,"眨眼",即闪烁

ON 和 OFF，甚至还有复制图案。

当然，这个游戏的哲学意义也很重要，因为它们表明，复杂的结构可以根据简单的规则演进，不必遵循任何一种预设的模式。

下面是该项目包含的一些主要概念：

- 利用 matplotlib imshow 来展示数据的二维网格；
- 利用 matplotlib 生成动画；
- 使用 numpy 数组；
- 将%运算符用于边界条件；
- 设置值的随机分布。

3.1 工作原理

因为生命游戏建立在 9 个方格的网格中，每个细胞有 8 个相邻细胞，如图 3-1 所示。模拟中的给定细胞(i, j)用二维数组[i][j]来存取，其中 i 和 j 分别是行和列的下标。在给定时间段，给定细胞的值取决于前一时间段它的邻居的状态。Conway 生命游戏有 4 个规则。

1. 如果一个细胞为 ON，邻居中少于两个为 ON，它变为 OFF。
2. 如果一个细胞为 ON，邻居中有两个或 3 个为 ON，它保持为 ON。
3. 如果一个细胞为 ON，邻居中超过 3 个为 ON，它变为 OFF。
4. 如果一个细胞为 OFF，邻居中恰好有 3 个为 ON，它变为 ON。

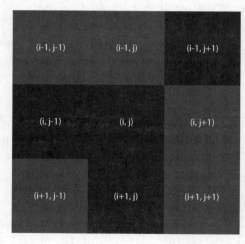

图 3-1　8 个相邻细胞

这些规则是为了反映一些基本方式，即一群生物体随时间推移的遭遇：如果细胞邻居少于 2 或多于 3，种群太少或种群太多都会杀死细胞，将细胞变为 OFF；如果种群平衡，细胞保持为 ON 并繁殖，将另一个细胞从 OFF 变为 ON。但是，在网格边缘的细胞呢？哪些细胞是自己的邻居？要回答这个问题，就要考虑边界条件，决定细胞在网格边缘或边界时的规则。我将采用环形边界条件，这意味着正方形网格卷起来，构成一个环面，从而解决这个问题。如图 3-2 所示，网格先卷起来，使它的水平边缘（A 和 B）相连，形成一个圆柱体，然后圆柱体的垂直边缘（C 和 D）相连，以形成一个环面。形成环面后，所有细胞都有邻居，因为整个空间没有边缘。

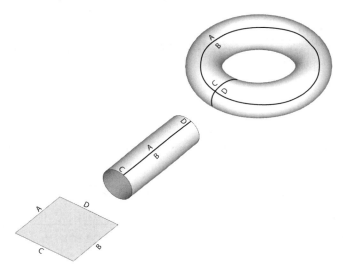

图 3-2　环型边界条件的概念视图

注意　　这类似于 Pac-Man（吃豆子）在边界的工作方式。如果超出了屏幕的顶部，就会重新在底部出现。如果超出了屏幕的左侧，就会重新在右侧出现。这种边界条件在二维模拟中很常见。

以下是算法描述，我们用它来应用这 4 个规则，并运行模拟。
1. 初始化网格中的细胞。
2. 在模拟的每个时间段，对于网格中每个细胞(i, j)，做下面的事：
   a. 根据它的邻居更新细胞(i, j)的值，同时考虑到边界条件；
   b. 更新网格值的显示。

3.2　所需模块

用 numpy 数组和 matplotlib 库来显示模拟的输出，用 matplotlib animation 模块更新模拟（对于 matplotlib 的综述见第 1 章）。

3.3 代码

我们将在 Python 解释器中一点一点地为模拟编写代码,考察不同部分所需的代码片段。要查看完整的项目代码,请直接跳到 3.4 节。

首先,导入该项目使用的模块:

```
>>> import numpy as np
>>> import matplotlib.pyplot as plt
>>> import matplotlib.animation as animation
```

现在让我们创建网格。

3.3.1 表示网格

为了在网格上表示细胞的活(ON)或死(OFF),分别用 255 和 0 作为 ON 和 OFF 的值。我们将采用 matplotlib 的 imshow()方法,来显示网格当前的状态,将一个数字矩阵表示为一张图像。请输入以下内容:

```
❶ >>> x = np.array([[0, 0, 255], [255, 255, 0], [0, 255, 0]])
❷ >>> plt.imshow(x, interpolation='nearest')
  plt.show()
```

在❶行,定义了 3×3 的二维 numpy 数组,数组中的每个元素是一个整数值。然后,在❷行,用 plt.show()方法将这个矩阵的值显示为图像,并给 interpolation 选项传入'nearest'值,以得到尖锐的边缘(否则是模糊的)。

图 3-3 展示了这段代码的输出。

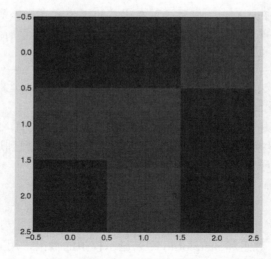

图 3-3　显示网格的值

注意，0（OFF）显示为暗灰色，255（ON）显示为浅灰色，这是 imshow()使用的默认颜色。

3.3.2　初始条件

要开始模拟，先为二维网格中的每个细胞设置一个初始状态。可以使用 ON 和 OFF 细胞的随机分布，看看会出现怎样的图案，或者添加一些特定的图案，看看它们如何发展。我们将探讨这两种方法。

要采用随机的初始状态，就使用 numpy 中 random 模拟的 choice()方法。输入以下内容：

```
np.random.choice([0, 255], 4*4, p=[0.1, 0.9]).reshape(4, 4)
```

下面是输出：

```
array([[255, 255, 255, 255],
       [255, 255, 255, 255],
       [255, 255, 255, 255],
       [255, 255, 255, 0]])
```

np.random.choice 从给定的列表[0,255]中选择一个值，每个值出现的概率由参数 p=[0.1, 0.9]指定。这里，你要求 0 出现的概率是 0.1（或 10%），255 出现的概率是 90%（p 中两个值相加必须等于 1）。因为这个 choice()方法创建了 16 个值的一维数组，所以用.reshape 使它成为一个二维数组。

要建立初始条件来匹配特定图案，而不是只填入一组随机值，就将二维网格初始化为零，然后用一个方法在网格的特定行和列增加一个图案，如下所示：

```
    def addGlider(i, j, grid):
        """adds a glider with top left cell at (i, j)"""
❶       glider = np.array([[0, 0, 255],
                          [255, 0, 255],
                          [0, 255, 255]])
❷       grid[i:i+3, j:j+3] = glider
❸   grid = np.zeros(N*N).reshape(N, N)
❹   addGlider(1, 1, grid)
```

在❶行，用 3×3 的 numpy 数组定义了滑翔机图案（看上去是一种在网格中平稳穿越的图案）。在❷行，可以看到如何用 numpy 的切片操作，将这种图案数组复制到模拟的二维网格中，它的左上角放在 i 和 j 指定的坐标。在❸行，创建 N×N 的零值数组，在❹行，调用 addGlider()方法，初始化带有滑翔机图案的网格。

3.3.3　边界条件

现在，我们可以想一下如何实现环形边界条件。首先，让我们看看在 N×N 网格的右边缘会发生什么情况。i 行最后一个细胞用 grid[i][N-1]来访问。它右侧的邻居是grid[i][N]，但根据环形边界条件，grid[i][N]访问的值应该由grid[i][0]代替。下面是一种实现方法：

```
if j == N-1:
    right = grid[i][0]
else:
    right = grid[i][j+1]
```

当然，需要在网格的左侧，顶部和底部应用类似的边界条件，但这样做要加入许多代码，因为需要检测网格的4个边缘。更简洁的方式是用Python的取模（%）运算符，如下所示：

```
>>> N = 16
>>> i1 = 14
>>> i2 = 15
>>> (i1+1)%N
15
>>> (i2+1)%N
0
```

如你所见，%运算符给出整数除以N的余数。可以用这个运算符让值在边缘折返，像这样重写网格访问代码：

```
right = grid[i][(j+1)%N]
```

现在，如果一个细胞在网格边缘（换言之，如果j= N-1），用这种方法请求右边的细胞就会得到(j +1) %N，这将j设回0，让网格右侧卷曲到左侧。如果在网格底部做同样的事，它就折返到顶部。

3.3.4 实现规则

生命游戏的规则基于相邻细胞的 ON 或 OFF 数目。为了简化这些规则的应用，可以计算出处于 ON 状态的相邻细胞总数。因为 ON 状态的值为 255，所以可以对所有相邻细胞的值求和，再除以 255，来获得 ON 细胞的数量。下面是相关的代码：

```
        # apply Conway's rules
        if grid[i, j] == ON:
❶           if (total < 2) or (total > 3):
                newGrid[i, j] = OFF
        else:
            if total == 3:
❷               newGrid[i, j] = ON
```

在❶行，如果少于 2 个相邻细胞为 ON 或多于 3 个相邻细胞为 ON，该细胞就由 ON 变成 OFF。❷行代码仅适用于 OFF 细胞：如果恰好有 3 个相邻细胞为 ON，该细胞就变成 ON。

现在该编写模拟的完整代码了。

3.3.5 向程序发送命令行参数

下面的代码向程序发送命令行参数:

```
# main() function
def main():
    # command line argumentss are in sys.argv[1], sys.argv[2], ...
    # sys.argv[0] is the script name and can be ignored
    # parse arguments
❶   parser = argparse.ArgumentParser(description="Runs Conway's Game of Life
        simulation.")
    # add arguments
❷   parser.add_argument('--grid-size', dest='N', required=False)
❸   parser.add_argument('--mov-file', dest='movfile', required=False)
❹   parser.add_argument('--interval', dest='interval', required=False)
❺   parser.add_argument('--glider', action='store_true', required=False)
    args = parser.parse_args()
```

main()函数首先定义了程序的命令行参数。在❶行，用 argparse 类为代码添加命令行选项，然后在接下来几行中添加各种选项。在❷行，指定模拟网格的大小 N。在❸行，指定保存.mov 文件的名称。在❹行，设置动画更新间隔的毫秒数。在❺行，用滑翔机图案开始模拟。

3.3.6 初始化模拟

继续看代码，接下来一段，对模拟初始化:

```
    # set grid size
    N = 100
    if args.N and int(args.N) > 8:
        N = int(args.N)

    # set animation update interval
    updateInterval = 50
    if args.interval:
        updateInterval = int(args.interval)

    # declare grid
❶   grid = np.array([])
    # check if "glider" demo flag is specified
    if args.glider:
        grid = np.zeros(N*N).reshape(N, N)
        addGlider(1, 1, grid)
    else:
        # populate grid with random on/off - more off than on
        grid = randomGrid(N)
```

仍在 main()函数内，命令行选项解析后，这部分代码应用命令行传入的所有参数。例如，❶行后的几行设置初始条件，要么是默认的随机图案，要么是滑翔机图案。最后，设置动画。

```
    # set up the animation
❶   fig, ax = plt.subplots()
    img = ax.imshow(grid, interpolation='nearest')
❷   ani = animation.FuncAnimation(fig, update, fargs=(img, grid, N, ),
                                  frames=10,
                                  interval=updateInterval,
                                  save_count=50)

    # number of frames?
    # set the output file
    if args.movfile:
        ani.save(args.movfile, fps=30, extra_args=['-vcodec', 'libx264'])

    plt.show()
```

在❶行，配置 matplotlib 的绘图和动画参数。在❷行，animation.FuncAnimation() 调用函数 update()，该函数在前面的程序中定义，根据 Conway 生命游戏的规则，采用环形边界条件来更新网格。

3.4 完整代码

下面是生命模拟游戏的完整程序。也可以从 https://github.com/electronut/pp/blob/master/conway/conway.py 下载该项目的代码。

```
import sys, argparse
import numpy as np
import matplotlib.pyplot as plt
import matplotlib.animation as animation

ON = 255
OFF = 0
vals = [ON, OFF]

def randomGrid(N):
    """returns a grid of NxN random values"""
    return np.random.choice(vals, N*N, p=[0.2, 0.8]).reshape(N, N)

def addGlider(i, j, grid):
    """adds a glider with top-left cell at (i, j)"""
    glider = np.array([[0, 0, 255],
                       [255, 0, 255],
                       [0, 255, 255]])
    grid[i:i+3, j:j+3] = glider

def update(frameNum, img, grid, N):
    # copy grid since we require 8 neighbors for calculation
    # and we go line by line
    newGrid = grid.copy()
    for i in range(N):
        for j in range(N):
            # compute 8-neghbor sum using toroidal boundary conditions
            # x and y wrap around so that the simulation
            # takes place on a toroidal surface
```

```python
            total = int((grid[i, (j-1)%N] + grid[i, (j+1)%N] +
                         grid[(i-1)%N, j] + grid[(i+1)%N, j] +
                         grid[(i-1)%N, (j-1)%N] + grid[(i-1)%N, (j+1)%N] +
                         grid[(i+1)%N, (j-1)%N] + grid[(i+1)%N, (j+1)%N])/255)
            # apply Conway's rules
            if grid[i, j] == ON:
                if (total < 2) or (total > 3):
                    newGrid[i, j] = OFF
            else:
                if total == 3:
                    newGrid[i, j] = ON

    # update data
    img.set_data(newGrid)
    grid[:] = newGrid[:]
    return img,

# main() function
def main():
    # command line arguments are in sys.argv[1], sys.argv[2], ...
    # sys.argv[0] is the script name and can be ignored
    # parse arguments
    parser = argparse.ArgumentParser(description="Runs Conway's Game of Life
        simulation.")
  # add arguments
    parser.add_argument('--grid-size', dest='N', required=False)
    parser.add_argument('--mov-file', dest='movfile', required=False)
    parser.add_argument('--interval', dest='interval', required=False)
    parser.add_argument('--glider', action='store_true', required=False)
    parser.add_argument('--gosper', action='store_true', required=False)
    args = parser.parse_args()

    # set grid size
    N = 100
    if args.N and int(args.N) > 8:
        N = int(args.N)

    # set animation update interval
    updateInterval = 50
    if args.interval:
        updateInterval = int(args.interval)

    # declare grid
    grid = np.array([])
    # check if "glider" demo flag is specified
    if args.glider:
        grid = np.zeros(N*N).reshape(N, N)
        addGlider(1, 1, grid)
    else:
        # populate grid with random on/off - more off than on
        grid = randomGrid(N)

    # set up the animation
    fig, ax = plt.subplots()
    img = ax.imshow(grid, interpolation='nearest')
    ani = animation.FuncAnimation(fig, update, fargs=(img, grid, N, ),
                                  frames=10,
                                  interval=updateInterval,
```

```
                          save_count=50)
    # number of frames?
    # set the output file
    if args.movfile:
        ani.save(args.movfile, fps=30, extra_args=['-vcodec', 'libx264'])

    plt.show()
# call main
if __name__ == '__main__':
    main()
```

3.5　运行模拟人生的游戏

现在运行代码：

```
$ python3 conway.py
```

这里采用模拟的默认参数：100×100 个细胞的网格，50 毫秒的更新间隔。观看模拟时，你会看到它如何进行，随着时间的推移创建并保持各种图案，如图 3-4 所示。

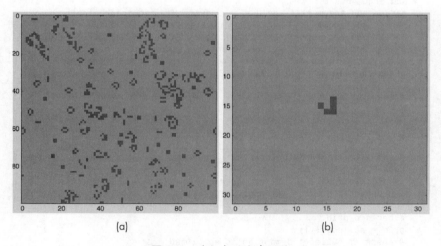

图 3-4　进行中的生命游戏

图 3-5 展示了模拟中可以寻找的一些图案。除了滑翔机，请寻找 3 细胞的闪光灯，以方块或面包的形状等静态图案。

现在，改变一下，用这些参数来运行模拟：

```
$ python conway.py --grid-size 32 --interval 500 --glider
```

这创建了 32×32 的模拟网格，每 500 毫秒更新动画，并采用初始的滑翔机图案，如图 3-5 的右下图所示。

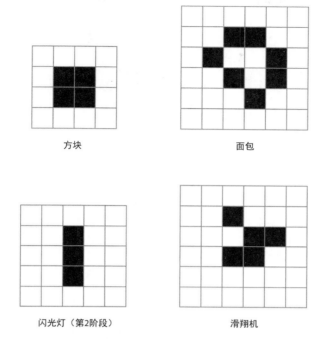

图 3-5　生命游戏中的图案

3.6　小结

这个项目探讨了 Conway 生命游戏。我们学习了如何基于某些规则，来建立基本的计算机模拟，以及如何用 matplotlib，随着系统的演进，让系统的状态可视化。

我们的 Conway 生命游戏实现更强调简单，而不是性能。有许多不同的方式可以加快生命游戏的计算，关于如何做到这一点，有大量的研究。快速在互联网搜索一下，会发现很多这样的研究。

3.7　实验

下面有一些方法，可以进一步试验 Conway 生命游戏。

1. 编写 addGosperGun() 方法，在网格中添加如图 3-6 所示的图案。这种图案被称为"高斯帕滑翔机枪（Gosper Glider Gun）"。运行模拟并观察枪的行为。

2. 编写 readPattern() 方法，从文本文件读取初始图案，并用它来设置模拟的初始条件。下面是该文件的建议格式：

```
8
0 0 0 255 ...
```

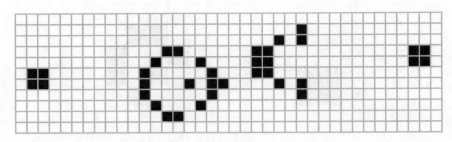

图 3-6　高斯帕滑翔机枪

该文件的第一行定义了 N，其余部分是 N×N 个整数（0 或 255），由空格隔开。你可以用 Python 的方法，如 open 和 file.read 来实现。这种探索有助于研究任何给定的图案在生命游戏规则下是如何演变的。添加命令行选项--pattern-file，在运行程序时使用此文件。

第 4 章
用 Karplus-Strong 算法产生音乐泛音

任何乐音的主要特点是它的音调，即频率。这是每秒的振动数，单位是赫兹（Hz）。例如，从原声吉他的第三根弦产生音符 D，频率是 146.83 赫兹。可以在计算机上创建频率为 146.83 Hz 的正弦波，来模拟这种声音，如图 4-1 所示。

遗憾的是，如果在计算机上播放这个正弦波，听起来不会像吉他或钢琴。播放相同的音符时，是什么让计算机的声音与乐器如此不同呢？

在吉他上拨弦时，乐器产生了不同强度的混合频率，如图 4-2 的频谱图所示。刚拨弦时声音最强，强度随着时间推移而逐渐消失。拨动吉他 D 弦听到的主要频率称为基本频率，是 146.83 赫兹，但也会听到该频率的一些倍数，称为泛音。任何乐器的声音都由这种基本频率和泛音组成，正是这种组合，让吉他听起来像吉他。

如你所见，在计算机上模拟拨弦乐器的声音，要能同时生成基本频率和泛音。诀窍是利用 Karplus-Strong 算法。

这个项目使用 Karplus-Strong 算法，产生 5 个类似吉他的音符，它们属于一个音阶（一系列相关的音符）。我们会让产生这些音符的算法可视化，并将声音保存为 WAV 文件。我们还会创建一种方式，随机演奏它们，并学习如何做到以下几点：

- 用 Python 的 deque 类实现环形缓冲区；
- 使用 numpy 数组和 ufuncs；
- 用 pygame 播放 WAV 文件；
- 用 matplotlib 绘图；
- 演奏五声音阶。

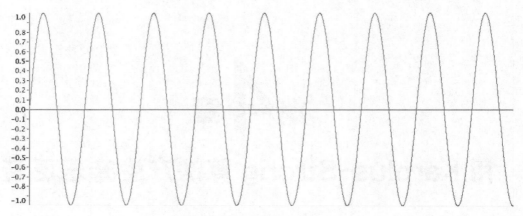

图 4-1　146.83 Hz 的正弦波

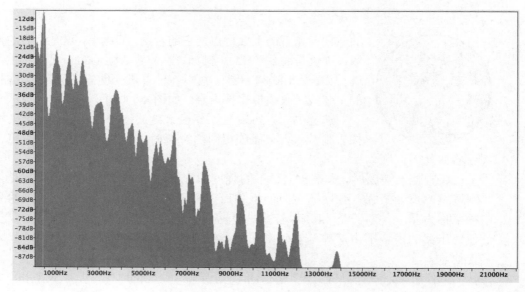

图 4-2　吉他上弹奏音符 D4 的频谱图

除了用 Python 实现 Karplus-Strong 算法，我们还会探讨 WAV 文件格式，看看如何产生五声音阶中的音符。

4.1 工作原理

利用偏移值构成的环形缓冲区,模拟两端固定的弦,类似于吉他弦,Karplus-Strong 算法可以模拟拨弦的声音。

环形缓冲区(也称为循环缓冲区)是一个固定长度的缓冲区(就是值的数组),但折返到本身。换言之,如果到达缓冲区的末尾,则下一个元素就是缓冲区的第一个元素(环形缓冲区的更多信息,参见 4.3.1 小节)。

根据公式 $N=S/f$,环形缓冲区的长度(N)与振动的基本频率有关,其中 S 是采样率,f 是频率。

模拟开始时,缓冲区中填充一些随机值,范围在[-0.5, 0.5],可以认为代表了拨弦在振动时的随机偏移。

除了环形缓冲区,可以用一个样本缓冲区,保存任何特定时间的声音强度。这个缓冲区的长度和采样率决定了声音片段的长度。

4.1.1 模拟

模拟持续进行,直到样本缓冲填充了一种反馈方案,如图 4-3 所示。对于模拟的每一步,做以下几件事:

1. 将环形缓冲区的第一个值存入样本缓冲区;
2. 计算环形缓冲区前两个元素的平均值;
3. 平均值乘以一个衰减系数(在这个例子中,是 0.995);
4. 将该值添加到环形缓冲区的末尾;
5. 移除环形缓冲区的第一个元素。

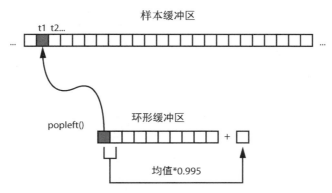

图 4-3 环形缓冲区和 Karplus-Strong 算法

为了模拟拨弦,在环形缓冲区填入一些数字,代表波的能量。样本缓冲区表示最终的声音数据，通过迭代遍历环形缓冲区中的值来创建。使用平均方案（等一下解释）来更新环形缓冲区的值。

这种反馈方案旨在模拟能量通过一根振动的弦。根据物理学，对于一根振动的弦，基本频率与它的长度成反比。因为我们感兴趣的是产生一定频率的声音，所以选择一个环形缓冲区，其长度与该频率成反比。模拟第 1 步的求平均值相当于"低通滤波器"，切断较高频率，并允许较低频率通过，从而消除高次谐波（即基频的较大倍数），因为我们的主要兴趣是基频。最后，用衰减因子来模拟波沿着弦来回传播的能量损失。

模拟第 1 步中用到的样本缓冲区，表示产生的声音随时间变化的振幅。为了计算任何给定的振幅，需要计算环形缓冲区前两个元素的平均值，结果乘以衰减系数，更新环形缓冲区。这个计算的值被添加到环形缓冲区的末尾，并删除第一个元素。

现在，我们来看看该算法执行的一个简单例子。下表表示两个连续时间步骤的环形缓冲区。环形缓冲区中的每个值表示声音的振幅。该缓冲区具有 5 个元素，它们开始填充了一些数字。

时间步骤 1	0.1	-0.2	0.3	0.6	-0.5
时间步骤 2	-0.2	0.3	0.6	-0.5	-0.199

从第 1 步转到第 2 步，像这样应用 Karplus-Strong 算法。第一行的第一个值 0.1 被删除，第 1 步中所有后续值以相同顺序加入第 2 行，它表示时间上的第 2 步。第 2 步的最后一个值是第 1 步的第一个值和最后一个值的衰减平均值，计算方法是 0.995×（(0.1+ –0.5)÷2）= –0.199。

4.1.2 创建 WAV 文件

波形音频文件格式（WAV）用于存储音频数据。这种格式对小的音乐项目比较方便，因为它简单，不需要处理复杂的压缩技术。

在最简单的形式中，WAV 文件由一系列比特构成，表示在给定时间点所记录的声音的振幅，称为分辨率。本项目中使用 16 位分辨率。WAV 文件也有一组采样率，是音频每秒采样或读取的次数。本项目采用 44100 赫兹，即 CD 中使用的采样率。让我们用 Python 产生 5 秒的 220 Hz 正弦波音频片段。首先，用这个公式表示正弦波：

$$A = \sin(2\pi ft)$$

这里，A 是声波的振幅，f 是频率，t 是当前时间的索引。现在，重写这个公式如下：

$$A = \sin(2\pi fi/R)$$

在这个公式中，i 是样本的索引，R 是采样率。用这两个方程，可以按如下方式创建一个 200 赫兹的正弦波 WAV 文件：

```
import numpy as np
import wave, math

sRate = 44100
nSamples = sRate * 5
❶ x = np.arange(nSamples)/float(sRate)
❷ vals = np.sin(2.0*math.pi*220.0*x)
❸ data = np.array(vals*32767, 'int16').tostring()
  file = wave.open('sine220.wav', 'wb')
❹ file.setparams((1, 2, sRate, nSamples, 'NONE', 'uncompressed'))
  file.writeframes(data)
  file.close()
```

在❶行和❷行，根据第二个正弦波方程，创建振幅值的 numpy 数组（见第 1 章）。对数组应用函数，如 sin() 函数，numpy 数组是快速、便捷的方式。

在❸行，计算好的、在[-1, 1]范围的正弦波值被放大为 16 位值，并转换成字符串，以便写入文件。在❹行，为 WAV 文件设置参数，在这个例子中，是单通道（单声道）、两字节（16 位）、无压缩的格式。图 4-4 展示了生成的 sine220.wav 文件在 Audacity 中的样子，Audacity 是一款免费的音频编辑器。正如预期的那样，我们看到了频率 220 赫兹的正弦波，如果播放该文件，会听到 5 秒的 220 Hz 音调。

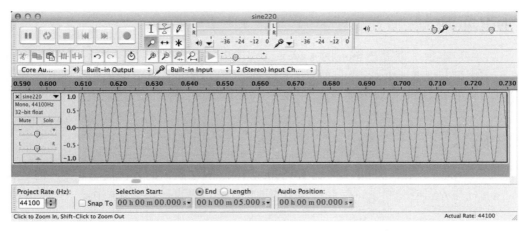

图 4-4　220 Hz 的正弦波

4.1.3　小调五声音阶

"音阶"是一系列升高或降低音高（频率）的音符。"音程"是两个音高之间的差。通常，一首音乐的所有音符都从特定音阶中选择。"半音"是音阶的基本单位，是西方音乐最小的音程。全音是半音长度的两倍。"大调音阶"是最常见的一种音阶，间隔模式是"全音-全音-半音-全音-全音-全音-半音"。

这里，我们将快速进入五声音阶，因为我们要产生这个音阶的音符。本节将告诉你，我们如何得出程序中用到的频率数字，利用 Karplus-Strong 算法来生成这些音符。"五声音阶"是五个音符的音阶。例如，美国著名的歌曲"Oh! Susanna"就是基于五声音阶的。这种音阶的一个变种是小调五声音阶。

这种音阶定义为音符序列"（全音+半音）-全音-全音-（全音+半音）-全音"。因此，C 小调五声音阶包含音符 C、降 E、F、G 和降 B。表 4-1 列出了小调五声音阶中 5 个音符的频率，我们将用 Karplus-Strong 算法来产生它（这里，C4 表示钢琴第 4 个八度的 C，或习惯称为中音 C）。

表 4-1　小调五声音阶的音符

音符	频率（Hz）
C4	261.6
降 E	311.1
F	349.2
G	392.0
降 B	466.2

4.2　所需模块

本项目将用 Python 的 wave 模块来创建 WAV 格式的音频文件。使用 numpy 数组来实现 Karplus-Strong 算法，用 Python 集合中的 deque 类来实现环形缓冲区。还会用 pygame 模块来播放 WAV 文件。

4.3　代码

现在，让我们开发实现 Karplus-Strong 算法所需的各种代码片段，然后将它们合并成完整的程序。要查看完整的项目代码，请直接跳到 4.4 节。

4.3.1　用 deque 实现环形缓冲区

回想一下，前面提到 Karplus-Strong 算法使用环形缓冲区来生成一个音符。我们将利用 Python 的 deque 容器（发音为"deck"，是 Python 的 collections 模块的一部分），来实现环形缓冲区，它用一个数组提供了专门的容器数据类型。可以从 deque 的开始（头）或末尾（尾）插入或删除元素（见图 4-5）。这种插入和删除过程是 O(1)或"常数时间"的操作，这意味着不论 deque 的容器变得多大，它需要的时间都是相同的。

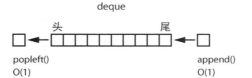

图 4-5 使用 deque 的环形缓冲区

下面的代码展示了如何在 Python 中使用 deque：

```
>>> from collections import deque
❶ >>> d = deque(range(10))
>>> print(d)
deque([0, 1, 2, 3, 4, 5, 6, 7, 8, 9])
❷ >>> d.append(-1)
>>> print(d)
deque([0, 1, 2, 3, 4, 5, 6, 7, 8, 9, -1])
❸ >>> d.popleft()
0
>>> print d
deque([1, 2, 3, 4, 5, 6, 7, 8, 9, -1])
```

在❶行，传入 range()方法创建的一个列表，创建了 deque 容器。在❷行，将一个元素添加到 deque 容器的末尾，在❸行，从 deque 的头部弹出（删除）第一个元素。这两个操作都很快。

4.3.2 实现 Karplus-Strong 算法

也可以用 deque 容器，针对环形缓冲区，实现 Karplus-Strong 算法，如下所示：

```
# generate note of given frequency
def generateNote(freq):
    nSamples = 44100
    sampleRate = 44100
    N = int(sampleRate/freq)
    # initialize ring buffer
❶   buf = deque([random.random() - 0.5 for i in range(N)])
    # initialize samples buffer
❷   samples = np.array([0]*nSamples, 'float32')
    for i in range(nSamples):
❸       samples[i] = buf[0]
❹       avg = 0.996*0.5*(buf[0] + buf[1])
        buf.append(avg)
        buf.popleft()

    # convert samples to 16-bit values and then to a string
    # the maximum value is 32767 for 16-bit
❺   samples = np.array(samples*32767, 'int16')
❻   return samples.tostring()
```

在❶行，用范围在[-0.5, 0.5]的随机数来初始化 deque。在❷行，建立一个浮点数组来保存声音采样。该数组的长度与采样率相符，这意味着将生成一秒钟的声音片段。

在❸行，deque 的第一个元素被复制到采样缓冲区。在❹行和随后几行中，可以看到低通滤波器和衰减在生效。在❺行，samples 数组的每个值乘以 32767（16 位带符号整数的取值范围是–32768～32,767），被转换成 16 位的格式。在❻行，它被转换成字符串表示形式，这是 wave 模块的要求，我们将用该模块，将这些数据保存到文件。

4.3.3 写 WAV 文件

拥有音频数据后，可以用 Python 的 wave 模块，将其写入 WAV 文件。

```
def writeWAVE(fname, data):
    # open file
❶   file = wave.open(fname, 'wb')
    # WAV file parameters
    nChannels = 1
    sampleWidth = 2
    frameRate = 44100
    nFrames = 44100
    # set parameters
❷   file.setparams((nChannels, sampleWidth, frameRate, nFrames,
                    'NONE', 'noncompressed'))
❸   file.writeframes(data)
    file.close()
```

在❶行，创建了一个 WAV 文件，在❷行，将参数设置为使用单声道、16 位、无压缩的格式。最后，在❸行，将数据写入文件。

4.3.4 用 pygame 播放 WAV 文件

现在，用 Python 的 pygame 模块播放算法生成的 WAV 文件。pygame 是用来编写游戏的流行 Python 模块。它基于简单直接媒体层（SDL）库，该库是高性能的底层库，让你能访问声音、图形，以及计算机的输入设备。

方便起见，我们将代码封装在 NotePlayer 类中，如下所示：

```
# play a WAV file
class NotePlayer:
    # constructor
    def __init__(self):
❶       pygame.mixer.pre_init(44100, -16, 1, 2048)
        pygame.init()
        # dictionary of notes
❷       self.notes = {}
    # add a note
    def add(self, fileName):
❸       self.notes[fileName] = pygame.mixer.Sound(fileName)
    # play a note
    def play(self, fileName):
        try:
❹           self.notes[fileName].play()
        except:
            print(fileName + ' not found!')
    def playRandom(self):
```

```
❺           """play a random note"""
            index = random.randint(0, len(self.notes)-1)
❻           note = list(self.notes.values())[index]
            note.play()
```

在❶行，预先初始化 pygame 的 mixer 类：44100 的采样率，16 位的有符号值，单声道，缓冲区大小为 2048。在❷行，创建一个音符的字典，它针对文件名保存 pygame 的声音对象。接下来，在 NotePlayer 的 add()方法中❸，创建声音对象，并将它保存在 notes 目录中。

请注意，在 play()中❹行如何使用目录，并播放与文件名相关联的声音对象。playRandom()方法从已生成的五个音符中随机选取一个播放。最后，在❺行，randint()从范围[0,4]中选择一个随机整数，在❻行从字典中挑选一个音符播放。

4.3.5 main()方法

现在看看 main()方法，它创建音符并处理各种命令行选项，以演奏音符。

```
    parser = argparse.ArgumentParser(description="Generating sounds with
        Karplus String Algorithm")
    # add arguments
❶   parser.add_argument('--display', action='store_true', required=False)
    parser.add_argument('--play', action='store_true', required=False)
    parser.add_argument('--piano', action='store_true', required=False)
    args = parser.parse_args()

    # show plot if flag set
    if args.display:
        gShowPlot = True
        plt.ion()

    # create note player
    nplayer = NotePlayer()

    print('creating notes...')
    for name, freq in list(pmNotes.items()):
        fileName = name + '.wav'
❷       if not os.path.exists(fileName) or args.display:
            data = generateNote(freq)
            print('creating ' + fileName + '...')
            writeWAVE(fileName, data)
        else:
            print('fileName already created. skipping...')

        # add note to player
❸       nplayer.add(name + '.wav')

        # play note if display flag set
        if args.display:
❹           nplayer.play(name + '.wav')
            time.sleep(0.5)

    # play a random tune
    if args.play:
        while True:
            try:
```

```
❺                    nplayer.playRandom()
                     # rest - 1 to 8 beats
❻                    rest = np.random.choice([1, 2, 4, 8], 1,
                                     p=[0.15, 0.7, 0.1, 0.05])
                     time.sleep(0.25*rest[0])
             except KeyboardInterrupt:
                 exit()
```

首先，用 argparse 为程序建立一些命令行选项，就像以前的项目中讨论的一样。在❶行，如果使用了--display 命令行选项，就用 matplotlib 绘图，显示 Karplus-Strong 算法中波形如何演变。ion()调用启动了 matplotlib 的交互模式。然后，创建 NotePlayer 类的一个实例，用 generateNote()方法生成五声音阶的音符。五个音符的频率在全局字典 pmNotes 中定义。

在❷行，利用 os.path.exists()方法来查看 WAV 文件是否已创建。如果已创建，就跳过计算（如果多次运行该程序，这是一个方便的优化）。

计算好音符、创建了 WAV 文件后，在❸行将该音符添加 NotePlayer 字典中，如果使用了--display 命令行选项，就在❹行播放它。

在❺行，如果使用了--play 选项，NotePlayer 的 playRandom()方法就从五个音符中随机播放一个音符。要让一个音符序列听起来更像音乐，需要在播放的音符之间添加休止符，所以在❻行，用 numpy 模块的 random.choice()方法，选择随机的休止时间。该方法还可以选择休止间隔的概率，设置之后可以让两拍的休止最有可能，八拍的休止最不可能。请试着改变这些值，以创建自己的音乐风格！

4.4 完整代码

现在将程序整合在一起。完整的代码如下，也可以从 https://github.com/electronut/pp/blob/master/karplus/ks.py 下载。

```python
import sys, os
import time, random
import wave, argparse, pygame
import numpy as np
from collections import deque
from matplotlib import pyplot as plt

# show plot of algorithm in action?
gShowPlot = False

# notes of a Pentatonic Minor scale
# piano C4-E(b)-F-G-B(b)-C5
pmNotes = {'C4': 262, 'Eb': 311, 'F': 349, 'G':391, 'Bb':466}

# write out WAV file
def writeWAVE(fname, data):
    # open file
    file = wave.open(fname, 'wb')
    # WAV file parameters
```

```python
        nChannels = 1
        sampleWidth = 2
        frameRate = 44100
        nFrames = 44100
        # set parameters
        file.setparams((nChannels, sampleWidth, frameRate, nFrames,
                        'NONE', 'noncompressed'))
        file.writeframes(data)
        file.close()

# generate note of given frequency
def generateNote(freq):
    nSamples = 44100
    sampleRate = 44100
    N = int(sampleRate/freq)
    # initialize ring buffer
    buf = deque([random.random() - 0.5 for i in range(N)])
    # plot of flag set
    if gShowPlot:
        axline, = plt.plot(buf)
    # initialize samples buffer
    samples = np.array([0]*nSamples, 'float32')
    for i in range(nSamples):
        samples[i] = buf[0]
        avg = 0.995*0.5*(buf[0] + buf[1])
        buf.append(avg)
        buf.popleft()
        # plot of flag set
        if gShowPlot:
            if i % 1000 == 0:
                axline.set_ydata(buf)
                plt.draw()

    # convert samples to 16-bit values and then to a string
    # the maximum value is 32767 for 16-bit
    samples = np.array(samples*32767, 'int16')
    return samples.tostring()

# play a WAV file
class NotePlayer:
    # constructor
    def __init__(self):
        pygame.mixer.pre_init(44100, -16, 1, 2048)
        pygame.init()
        # dictionary of notes
        self.notes = {}
    # add a note
    def add(self, fileName):
        self.notes[fileName] = pygame.mixer.Sound(fileName)
    # play a note
    def play(self, fileName):
        try:
            self.notes[fileName].play()
        except:
            print(fileName + ' not found!')
    def playRandom(self):
        """play a random note"""
        index = random.randint(0, len(self.notes)-1)
```

```python
            note = list(self.notes.values())[index]
            note.play()

# main() function
def main():
    # declare global var
    global gShowPlot

    parser = argparse.ArgumentParser(description="Generating sounds with
        Karplus String Algorithm")
    # add arguments
    parser.add_argument('--display', action='store_true', required=False)
    parser.add_argument('--play', action='store_true', required=False)
    parser.add_argument('--piano', action='store_true', required=False)
    args = parser.parse_args()

    # show plot if flag set
    if args.display:
        gShowPlot = True
        plt.ion()

    # create note player
    nplayer = NotePlayer()

    print('creating notes...')
    for name, freq in list(pmNotes.items()):
        fileName = name + '.wav'
        if not os.path.exists(fileName) or args.display:
            data = generateNote(freq)
            print('creating ' + fileName + '...')
            writeWAVE(fileName, data)
        else:
            print('fileName already created. skipping...')

        # add note to player
        nplayer.add(name + '.wav')

        # play note if display flag set
        if args.display:
            nplayer.play(name + '.wav')
            time.sleep(0.5)

    # play a random tune
    if args.play:
        while True:
            try:
                nplayer.playRandom()
                # rest - 1 to 8 beats
                rest = np.random.choice([1, 2, 4, 8], 1,
                                        p=[0.15, 0.7, 0.1, 0.05])
                time.sleep(0.25*rest[0])
            except KeyboardInterrupt:
                exit()

    # random piano mode
    if args.piano:
        while True:
            for event in pygame.event.get():
```

```
            if (event.type == pygame.KEYUP):
                print("key pressed")
                nplayer.playRandom()
                time.sleep(0.5)
# call main
if __name__ == '__main__':
    main()
```

4.5 运行拨弦模拟

要运行这个项目的代码,就在命令行输入:

```
$ python3 ks.py –display
```

matplotlib 绘图展示了 Karplus-Strong 算法如何转换初始随机偏移,创建所需频率的波,如图 4-6 所示。

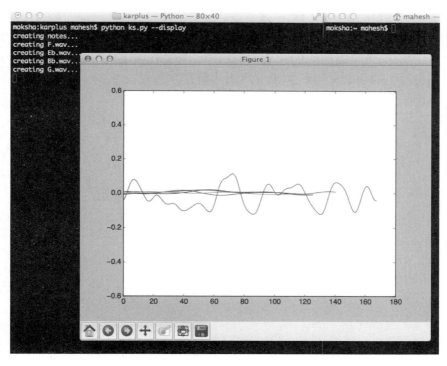

图 4-6　Karplus-Strong 算法的运行示例

现在尝试用这个程序播放随机音符。

```
$ python ks.py –play
```

这应该会用生成的五声音阶 WAV 文件播放随机的音符序列。

4.6 小结

这个项目用 Karplus-Strong 算法来模拟拨弦的声音，利用生成的 WAV 文件来播放音符。

4.7 实验

这里有一些值得试验的想法：

1. 用你在本章学到的技术创建一种方法，重现两根不同频率的弦一起振动发出的声音。记住，Karplus-Strong 算法产生的声音振幅可以叠加（放大到 16 位值并生成 WAV 文件之前）。现在，在第一次拨弦和第二次拨弦之间增加一点延时。

2. 写一个方法，从文本文件中读取音乐，并生成音符。然后用这些音符播放音乐。使用的格式可以是音符名称后面跟上整数的休止时间，就像这样：C4 1 F4 2 G4 1……

3. 为项目添加 --piano 命令行选项。如果程序用这个选项运行，用户应该能够按键盘上的 A、S、D、F 和 G 键，播放五个音符（提示：用 pygame.event.get 和 pygame.event.type）。

第5章
类鸟群：仿真鸟群

仔细观察一群鸟或一群鱼，你会发现，虽然群体由个体生物组成，但该群体作为一个整体似乎有它自己的生命。鸟群中的鸟在移动、飞越和绕过障碍物时，彼此之间相互定位。受到打扰或惊吓时会破坏编队，但随后重新集结，仿佛被某种更大的力量控制。

1986年，Craig Reynolds创造鸟类群体行为的一种逼真模拟，称为"类鸟群（Boids）"模型。关于类鸟群模型，值得注意的是，只有3个简单的规则控制着群体中个体间的相互作用，但该模型产生的行为类似于真正的鸟群。类鸟群模型被广泛研究，甚至被用来制作群体的计算机动画，如电影"蝙蝠侠归来（1992）"中的行军企鹅。

本项目将利用Reynolds的3个规则，创建一个类鸟群，模拟N只鸟的群体行为，并画出随着时间的推移，它们的位置和运动方向。我们还会提供一个方法，向鸟群中添加一只鸟，以及一种驱散效果，可以用于研究群体的局部干扰效果。类鸟群被称为"N体模拟"，因为它模拟了N个粒子的动态系统，彼此之间施加作用力。

5.1 工作原理

模拟类鸟群的三大核心规则如下：

分离：保持类鸟个体之间的最小距离；

列队：让每个类鸟个体指向其局部同伴的平均移动方向；

内聚：让每个类鸟个体朝其局部同伴的质量中心移动。

类鸟群模拟也可以添加其他规则，如避开障碍物，或受到打扰时驱散鸟群，在随后的小节中我们将会探讨这些。这个版本的类鸟群在模拟的每一步中，实现了这些核心规则。

- 对于群体中的所有类鸟个体，做以下几件事：
  - 应用三大核心规则；
  - 应用所有附加规则；
  - 应用所有边界条件。
- 更新类鸟个体的位置和速度。
- 绘制新的位置和速度。

如你所见，这些简单的规则创造了一个鸟群，它具有演变的复杂行为。

5.2 所需模块

下面是该模拟要用到的 Python 工具：

- numpy 数组，用于保存类鸟群的位置和速度；
- matplotlib 库，用于生成类鸟群动画；
- argparse，用于处理命令行选项；
- scipy.spatial.distance 模块，包含一些非常简洁的方法，计算点之间的距离。

注意　对类鸟群项目，我选择使用 matplotlib，是为了简单和方便。要尽可能快地绘制数量庞大的类鸟群，可能会使用类似 OpenGL 这样的库。本书的第三部分将详细探讨图形。

5.3 代码

首先，要计算类鸟群的位置和速度。接下来，要为模拟设置边界条件，看看如何绘制类鸟群，并实现前面讨论的类鸟群模拟规则。最后，我们会为模拟添加一些有趣的事件，即添加一些类鸟个体和驱散类鸟群。要查看完整的项目代码，请直接跳到 5.4 节。

5.3.1 计算类鸟群的位置和速度

类鸟群仿真需要从 numpy 数组取得信息,计算每一步中类鸟群个体的位置和速度。模拟开始时,将所有类鸟群个体大致放在屏幕中央,速度设置为随机的方向。

```
❶ import math
❷ import numpy as np

❸ width, height = 640, 480

❹ pos = [width/2.0, height/2.0] + 10*np.random.rand(2*N).reshape(N, 2)
❺ angles = 2*math.pi*np.random.rand(N)
❻ vel = np.array(list(zip(np.sin(angles), np.cos(angles))))
```

开始在❶行导入 math 模块,用于接下来的计算。在❷行,将 numpy 库导入为 np(少一些录入)。然后,设置屏幕上模拟窗口的宽度和高度❸。在❹行,创建一个 numpy 数组 pos,对窗口中心加上 10 个单位以内的随机偏移。代码 np.random.rand(2 * N) 创建了一个一维数组,包含范围在[0, 1]的 2N 个随机数。然后 reshape() 调用将它转换成二维数组的形状(N, 2),它将用于保存类鸟群个体的位置。也要注意,numpy 的广播规则在这里生效:1×2 的数组加到 N×2 的数组的每个元素上。

接下来,用以下方法,创建随机单位速度矢量数组(这些都是模为 1.0 的矢量,指向随机的方向):给定一个角度 t,数字对(cos(t), sin(t))位于半径为 1.0 的圆上,中心在原点(0, 0)。如果从原点到圆上的一点画一条线,就得到一个单位矢量,它取决于角度 A。如果随机选择角度 A,就得到一个随机速度矢量。图 5-1 展示了这个方案。

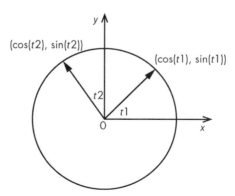

图 5-1 随机生成单位速度矢量

在❺行,生成一个数组,包含 N 个随机角度,范围在[0, 2pi],在❻行,用前面讨论的随机向量方法生成一个数组,并用内置的 zip() 方法将坐标分组。这里有 zip() 的一个简单例子。它将两个列表合并成一个元组的列表。

```
>>> zip([0, 1, 2], [3, 4, 5])
[(0, 3), (1, 4), (2, 5)]
```

这里，生成了两个数组，一个包含随机的位置，聚集在屏幕中心 10 像素的半径范围内，另一个包含随机方向的单位速度。这意味着在模拟开始时，类鸟群盘旋在屏幕中心，指向随机的方向。

5.3.2 设置边界条件

鸟儿飞翔在无际的天空，但类鸟群必须在有限的空间中运动。要创建这个空间，就要创建边界条件，就像第 3 章中为 Conway 模拟创建环形边界条件一样。在这个例子中，我们采用"平铺小块边界条件"（实际上是第 3 章中使用的边界条件的连续空间版本）。

将类鸟群模拟想象成发生在一个平铺的空间：如果类鸟群个体离开一个小块，它将从相反的方向进入到相同的小块。环形边界条件和小块边界条件之间的主要区别是，类鸟群模拟不会发生在离散的网格上，而是在一个连续区域移动。图 5-2 展示了这些小块边界条件的样子。请看图 5-2 中间的小块。飞出右侧的鸟儿正进入右边的小块，但该边界条件确保它们实际上通过平铺在左边的小块，又回到了中心的小块。在顶部和底部的小块，可以看到同样的事情发生。

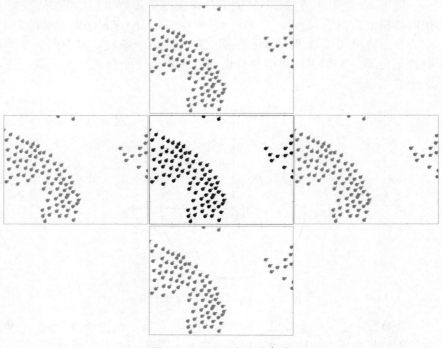

图 5-2　平铺小块边界条件

下面是如何为类鸟群模拟实现平铺小块边界条件：

```
    def applyBC(self):
        """apply boundary conditions"""
        deltaR = 2.0
        for coord in self.pos:
❶           if coord[0] > width + deltaR:
                coord[0] = - deltaR
            if coord[0] < - deltaR:
                coord[0] = width + deltaR
            if coord[1] > height + deltaR:
                coord[1] = - deltaR
            if coord[1] < - deltaR:
                coord[1] = height + deltaR
```

在❶行，如果 x 坐标比小块的宽度大，则将它设置回小块的左侧边缘。该行中的 deltaR 提供了一个微小的缓冲区，它允许类鸟群个体开始从相反方向回来之前，稍稍移出小块之外一点，从而产生更好的视觉效果。在小块的左侧、顶部和底部边缘执行类似的检查。

5.3.3 绘制类鸟群

要生成动画，需要知道类鸟群个体的位置和速度，并有办法在每个时间步骤中表示位置和运动方向。

1. 绘制类鸟群个体的身体和头部

为了生成类鸟群动画，我们用 matplotlib 和一点小技巧来绘制位置和速度。将每个类鸟群个体画成两个圆，如图 5-3 所示。较大的圆代表身体，较小的圆表示头部。点 P 是身体的中心，H 是头部的中心。根据公式 $H = P + k \times V$ 来计算 H 的位置，其中 V 是类鸟群个体的速度，k 是常数。在任何给定时间，类鸟群个体的头指向运动的方向。这指明了类鸟群个体的移动方向，比只画身体更好。

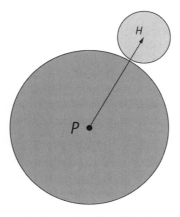

图 5-3　表示类鸟群个体

在下面的代码片段中，利用 matplotlib，用圆形标记画出类鸟群个体的身体。

```
fig = plt.figure()
ax = plt.axes(xlim=(0, width), ylim=(0, height))
```
❶ `pts, = ax.plot([], [], markersize=10, c='k', marker='o', ls='None')`
❷ `beak, = ax.plot([], [], markersize=4, c='r', marker='o', ls='None')`
❸ `anim = animation.FuncAnimation(fig, tick, fargs=(pts, beak, boids),`
`                                interval=50)`

在❶和❷行分别为类鸟群个体的身体（pts）和头部（beak）标记设置大小和形状。在❸行为动画窗口添加鼠标按钮事件。既然知道了如何绘制身体和喙，让我们看看如何更新它们的位置。

2. 更新类鸟群个体的位置

动画开始后，需要更新身体和头的位置，它指明了类鸟群个体移动的方向。用以下代码来实现：

❶ `vec = self.pos + 10*self.vel/self.maxVel`
❷ `beak.set_data(vec.reshape(2*self.N)[::2], vec.reshape(2*self.N)[1::2])`

在❶行，计算头部的位置，即在速度（vel）的方向上增加 10 个单位的位移。该位移确定了喙和身体之间的距离。在❷行，用头部位置的新值来更新（reshape）matplotlib 的轴（set_data）。[::2]从速度列表中选出偶数元素（x 轴的值），[1::2]选出奇数元素（Y 轴的值）。

5.3.4 应用类鸟群规则

现在，要在 Python 中实现类鸟群的 3 个规则。我们用"numpy 的方式"来完成这件事，避免循环并利用高度优化的 numpy 方法。

```
import numpy as np
from scipy.spatial.distance import squareform, pdist, cdist

    def test2(pos, radius):
        # get distance matrix
❶       distMatrix = squareform(pdist(pos))
        # apply threshold
❷       D = distMatrix < radius
        # compute velocity
❸       vel = pos*D.sum(axis=1).reshape(N, 1) - D.dot(pos)
        return vel
```

在❶行，用 squareform()和 pdist()方法（在 scipy 库中定义），来计算一组点之间两两的距离（从数组中任意取两点，计算距离，然后针对所有可能的两点对这么做）。例如，在下面代码中，有 3 个点，这意味着 3 种可能两点对：

```
>>> import numpy as np
>>> from scipy.spatial.distance import squareform, pdist
>>> x = np.array([[0.0, 0.0], [1.0, 1.0], [2.0, 2.0]])
```

```
>>> squareform(pdist(x))
array([[ 0.        ,  1.41421356,  2.82842712],
       [ 1.41421356,  0.        ,  1.41421356],
       [ 2.82842712,  1.41421356,  0.        ]])
```

squareform()方法给出一个 3×3 矩阵，其中项 M_{ij} 给出了点 P_i 和 P_j 之间的距离。接下来，在❷行，基于距离筛选这个矩阵。使用同样的 3 点的例子，得到如下结果：

```
>>> squareform(pdist(x)) < 1.4
array([[ True, False, False],
       [False,  True, False],
       [False, False,  True]], dtype=bool)
```

"<" 比较针对距离小于给定阈值（这个例子中是 1.4）的所有距离对，设置矩阵的项。这种紧凑的方式表达了你想要的结果，更接近实际的思考方式。

在❸行的方法有点复杂。D.sum()方法按列对矩阵中的 True 值求和。reshape 是必需的，因为和是 N 个值的一维数组（形如(N,)），而你希望它形如（N，1），这样它就能够与位置数组相乘。D.dot()就是矩阵和位置矢量的点积（乘法）。

test2 比 test1 的小得多，但它真正的优势是速度。让我们用 Python timeit 模块来比较前面两种方法的性能。下面的代码写在 Python 解释器中，假设函数 test1 和 test2 的代码在同一目录下的 test.py 文件中：

```
>>> from timeit import timeit
>>> timeit('test1(pos, 100)', 'from test import test1, N, pos, width, height',
number=100)
7.880876064300537
>>> timeit('test2(pos, 100)', 'from test import test2, N, pos, width, height',
number=100)
0.036969900131225586
```

在我的计算机上，没有循环的 numpy 代码比使用显式循环的代码快约 200 倍！但为什么呢？它们不是差不多都是一回事吗？

原因在于，作为一种解释型语言，Python 天生就比 C 语言这样的编译语言慢。numpy 库提供了高度优化的数组操作方法，带来了 Python 的方便和几乎 C 语言相等的性能（你会发现，如果重新组织算法，每次操作整个数组，不要循环单个元素来进行计算，这样 numpy 的效果最好）。

下面的方法利用前面讨论的 numpy 技术，应用类鸟群的 3 个规则：

```
    def applyRules(self):
        # apply rule #1: Separation
        D = distMatrix < 25.0
❶       vel = self.pos*D.sum(axis=1).reshape(self.N, 1) - D.dot(self.pos)
❷       self.limit(vel, self.maxRuleVel)

        # distance threshold for alignment (different from separation)
        D = distMatrix < 50.0

        # apply rule #2: Alignment
❸       vel2 = D.dot(self.vel)
```

```
            self.limit(vel2, self.maxRuleVel)
            vel += vel2;

            # apply rule #3: Cohesion
❹           vel3 = D.dot(self.pos) - self.pos
            self.limit(vel3, self.maxRuleVel)
            vel += vel3

            return vel
```

在❶行应用分离规则时，每个个体都被"推离"相邻个体一定距离，正如本节开始时的讨论。在❷行，计算出的速度被限制在某个最大值以内（没有这项检查，值将随着每个时间步骤增加，模拟将失控）。

在❸行应用列队规则时，50 个单位的半径内，所有相邻个体的速度之和限制为一个最大值。这样做是为了计算的最后速度不会无限增加。因此，任何给定的个体，都会受到指定半径内个体的平均速度影响，并据此列队（利用紧凑的 numpy 语法做这种计算，让事情变得简单、快捷。）

最后，在❹行应用内聚规则，为每个个体增加一个速度矢量，它指向一定半径内相邻个体的重心或几何中心。利用布尔距离矩阵和 numpy 的方法，实现紧凑的语法。

5.3.5 添加个体

类鸟群模拟的核心规则会导致类鸟群展示出群聚行为。但是，让我们在模拟过程中添加一个个体，看看表现如何，让事情变得更有趣。

下面的代码创建一个鼠标事件，让你点击鼠标左键添加一个个体。个体将出现在光标的位置，具有随机指定的速度。

```
    # add a "button press" event handler
❶   cid = fig.canvas.mpl_connect('button_press_event', buttonPress)
```

在❶行，用 mpl_connect()方法向 matplotlib 画布添加一个按钮按下事件。每次在模拟窗口按下鼠标时，buttonPress()方法都会被调用。

现在，为了处理鼠标事件，实际创建类鸟群个体，添加以下代码：

```
    def buttonPress(self, event):
        """event handler for matplotlib button presses"""
        # left-click to add a boid
❶       if event.button is 1:
❷           self.pos = np.concatenate((self.pos,
                                 np.array([[event.xdata, event.ydata]])),
                                 axis=0)
            # generate a random velocity
❸           angles = 2*math.pi*np.random.rand(1)
            v = np.array(list(zip(np.sin(angles), np.cos(angles))))
            self.vel = np.concatenate((self.vel, v), axis=0)
            self.N += 1
```

在❶行，确保鼠标事件是左键点击。在❷行，将(event.xdata, event.ydata)给出的鼠标位置添加到类鸟群的位置数组。在❸行和随后的行中，将一个随机速度矢量添加到类鸟群的速度数组，并将类鸟群的计数增加 1。

5.3.6 驱散类鸟群

3 个模拟规则保持类鸟群在移动时成为一个群体。但是，群体受到惊扰时，会发生什么？为了模拟这种情况，可以引入一种"驱散"效果：如果在用户界面（UI）窗口中单击右键，群体就会分散。你可以认为这是群体面对突然出现的捕食者的反应，或突然出现一声巨响惊吓了鸟群。下面是实现该效果的一种方式，它作为 buttonPress()方法的延续：

```
        # right-click to scatter boids
❶       elif event.button is 3:
            # add scattering velocity
❷           self.vel += 0.1*(self.pos - np.array([[event.xdata, event.ydata]]))
```

在❶行，检查鼠标按键是否是右键单击事件。在❷行，改变每个个体的速度，在干扰出现的点（即点击鼠标的位置）的相反的方向上增加一个分量。最初，类鸟群将飞离该点，但你会看到，3 个规则胜出，类鸟群将作为群体再次会聚。

5.3.7 命令行参数

下面是类鸟群程序如何处理命令行参数：

```
❶   parser = argparse.ArgumentParser(description="Implementing Craig
                                    Reynolds's Boids...")
    # add arguments
    parser.add_argument('--num-boids', dest='N', required=False)
    args = parser.parse_args()

    # set the initial number of boids
    N = 100
    if args.N:
        N = int(args.N)

    # create boids
    boids = Boids(N)
```

main()方法首先在❶行设置了一个命令行选项，使用我们熟悉的 argparse 模块。

5.3.8 Boids 类

接下来看看 Boids 类，它代表了模拟。

```
class Boids:
    """class that represents Boids simulation"""
    def __init__(self, N):
        """initialize the Boid simulation"""
        # initial position and velocities
```

第 5 章 类鸟群：仿真鸟群 71

```
❶          self.pos = [width/2.0, height/2.0] + 10*np.random.rand(2*N).reshape(N, 2)
           # normalized random velocities
           angles = 2*math.pi*np.random.rand(N)
           self.vel = np.array(list(zip(np.sin(angles), np.cos(angles))))
           self.N = N
           # minimum distance of approach
           self.minDist = 25.0
           # maximum magnitude of velocities calculated by "rules"
           self.maxRuleVel = 0.03
           # maximum magnitude of the final velocity
           self.maxVel = 2.0
```

Boid 类处理初始化，更新动画，并应用规则。在❶行和随后的行中，初始化位置和速度数组。

boids.tick()在每个时间步骤被调用，以便更新动画，如下所示：

```
def tick(frameNum, pts, beak, boids):
    #print frameNum
    """update function for animation"""
    boids.tick(frameNum, pts, beak)
    return pts, beak
```

我们还需要一种方法来限制某些矢量的值。否则，速度将在每个时间步骤无限制地增加，模拟将崩溃。

```
    def limitVec(self, vec, maxVal):
        """limit the magnitude of the 2D vector"""
        mag = norm(vec)
        if mag > maxVal:
            vec[0], vec[1] = vec[0]*maxVal/mag, vec[1]*maxVal/mag
❶   def limit(self, X, maxVal):
        """limit the magnitude of 2D vectors in array X to maxValue"""
        for vec in X:
            self.limitVec(vec, maxVal)
```

在❶行定义了 limit()方法，限制了数组中的值，采用模拟规则计算出的值。

5.4 完整代码

下面是类鸟群模拟的完整程序。也可以从 https://github.com/electronut/pp/blob/master/boids/boids.py 下载该项目的代码。

```
import sys, argparse
import math
import numpy as np
import matplotlib.pyplot as plt
import matplotlib.animation as animation
from scipy.spatial.distance import squareform, pdist, cdist
from numpy.linalg import norm

width, height = 640, 480

class Boids:
```

```python
"""class that represents Boids simulation"""
def __init__(self, N):
    """initialize the Boid simulation"""
    # initial position and velocities
    self.pos = [width/2.0, height/2.0] + 10*np.random.rand(2*N).reshape(N, 2)
    # normalized random velocities
    angles = 2*math.pi*np.random.rand(N)
    self.vel = np.array(list(zip(np.sin(angles), np.cos(angles))))
    self.N = N
    # minimum distance of approach
    self.minDist = 25.0
    # maximum magnitude of velocities calculated by "rules"
    self.maxRuleVel = 0.03
    # maximum maginitude of the final velocity
    self.maxVel = 2.0

def tick(self, frameNum, pts, beak):
    """Update the simulation by one time step."""
    # get pairwise distances
    self.distMatrix = squareform(pdist(self.pos))
    # apply rules:
    self.vel += self.applyRules()
    self.limit(self.vel, self.maxVel)
    self.pos += self.vel
    self.applyBC()
    # update data
    pts.set_data(self.pos.reshape(2*self.N)[::2],
                 self.pos.reshape(2*self.N)[1::2])
    vec = self.pos + 10*self.vel/self.maxVel
    beak.set_data(vec.reshape(2*self.N)[::2],
                  vec.reshape(2*self.N)[1::2])

def limitVec(self, vec, maxVal):
    """limit the magnitide of the 2D vector"""
    mag = norm(vec)
    if mag > maxVal:
        vec[0], vec[1] = vec[0]*maxVal/mag, vec[1]*maxVal/mag

def limit(self, X, maxVal):
    """limit the magnitide of 2D vectors in array X to maxValue"""
    for vec in X:
        self.limitVec(vec, maxVal)

def applyBC(self):
    """apply boundary conditions"""
    deltaR = 2.0
    for coord in self.pos:
        if coord[0] > width + deltaR:
            coord[0] = - deltaR
        if coord[0] < - deltaR:
            coord[0] = width + deltaR
        if coord[1] > height + deltaR:
            coord[1] = - deltaR
        if coord[1] < - deltaR:
            coord[1] = height + deltaR

def applyRules(self):
```

```python
            # apply rule #1: Separation
            D = self.distMatrix < 25.0
            vel = self.pos*D.sum(axis=1).reshape(self.N, 1) - D.dot(self.pos)
            self.limit(vel, self.maxRuleVel)

            # distance threshold for alignment (different from separation)
            D = self.distMatrix < 50.0

            # apply rule #2: Alignment
            vel2 = D.dot(self.vel)
            self.limit(vel2, self.maxRuleVel)
            vel += vel2;

            # apply rule #3: Cohesion
            vel3 = D.dot(self.pos) - self.pos
            self.limit(vel3, self.maxRuleVel)
            vel += vel3

            return vel

    def buttonPress(self, event):
        """event handler for matplotlib button presses"""
        # left-click to add a boid
        if event.button is 1:
            self.pos = np.concatenate((self.pos,
                                       np.array([[event.xdata, event.ydata]])),
                                      axis=0)
            # generate a random velocity
            angles = 2*math.pi*np.random.rand(1)
            v = np.array(list(zip(np.sin(angles), np.cos(angles))))
            self.vel = np.concatenate((self.vel, v), axis=0)
            self.N += 1
        # right-click to scatter boids
        elif event.button is 3:
            # add scattering velocity
            self.vel += 0.1*(self.pos - np.array([[event.xdata, event.ydata]]))

def tick(frameNum, pts, beak, boids):
    #print frameNum
    """update function for animation"""
    boids.tick(frameNum, pts, beak)
    return pts, beak

# main() function
def main():
    # use sys.argv if needed
    print('starting boids...')

    parser = argparse.ArgumentParser(description="Implementing Craig
                                    Reynold's Boids...")
    # add arguments
    parser.add_argument('--num-boids', dest='N', required=False)
    args = parser.parse_args()

    # set the initial number of boids
    N = 100
    if args.N:
```

```
        N = int(args.N)

    # create boids
    boids = Boids(N)

    # set up plot
    fig = plt.figure()
    ax = plt.axes(xlim=(0, width), ylim=(0, height))

    pts, = ax.plot([], [], markersize=10, c='k', marker='o', ls='None')
    beak, = ax.plot([], [], markersize=4, c='r', marker='o', ls='None')
    anim = animation.FuncAnimation(fig, tick, fargs=(pts, beak, boids),
                                   interval=50)

    # add a "button press" event handler
    cid = fig.canvas.mpl_connect('button_press_event', boids.buttonPress)

    plt.show()

# call main
if __name__ == '__main__':
    main()
```

5.5 运行类鸟群模拟

让我们来看看，运行模拟时会发生什么。输入以下内容：

```
$ python3 boids.py
```

类鸟群模拟开始，所有个体应该聚集在窗口的中心附近。让模拟运行一段时间，类鸟群应该开群聚，构成的模式类似于图 5-4 所示。

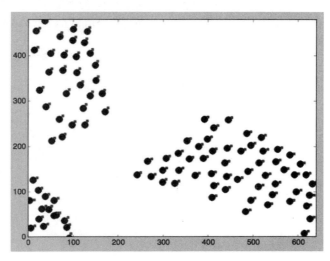

图 5-4　类鸟群的运行示例

点击模拟窗口。新的个体应该出现在该位置，当它遇到鸟群时，速度应改变。现在，单击鼠标右键。鸟群应首先分散，但随后重新聚集。

5.6 小结

这个项目利用 Craig Reynolds 提出的 3 个规则，模拟鸟群（或类鸟群）的聚集。我们学习了如何使用 numpy 数组，如何使用显式循环，以及用整个数组上的 numpy 方法来提高计算速度。我们利用 scipy.spatial 模块来执行快速和方便的距离计算，实现了一个 matplotlib 技巧，利用两个记号来表示个体的位置和方向。最后，增加了 UI 交互，可以按下鼠标按钮来改变 matplotlib 的绘图。

5.7 实验

下面有一些方式，可以进一步探索群聚行为。

1. 为类鸟群实现避障，编写一个新方法 avoidObstacle()，并在应用 3 个规则之后应用它，像下面这样：

```
self.vel += self.applyRules()
self.vel += self.avoidObstacle()
```

avoidObstacle()方法应该使用预先定义的元组(x, y, R)，为个体增加一个速度分量，将它推离障碍物的位置(x, y)，但只在个体处于障碍物的半径 R 之内时。可以将其视为是个体看见障碍物，并避开它的距离。可以用命令行选项指定(x, y, R)元组。

2. 如果类鸟群飞过一阵强风，会发生什么情况？以随机选择的时间步骤，对所有个体增加一个全局的速度分量，来模拟这种情况。类鸟群应该暂时受到风的影响，但风停止后又回到群聚状态。

第三部分

图片之乐

"你可以通过视觉观察到很多东西。"
——Yogi Berra

第6章
ASCII 文本图形

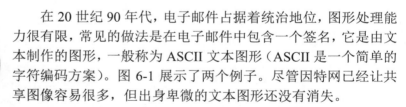

在 20 世纪 90 年代，电子邮件占据着统治地位，图形处理能力很有限，常见的做法是在电子邮件中包含一个签名，它是由文本制作的图形，一般称为 ASCII 文本图形（ASCII 是一个简单的字符编码方案）。图 6-1 展示了两个例子。尽管因特网已经让共享图像容易很多，但出身卑微的文本图形还没有消失。

ASCII 文本图形的源头是 19 世纪后期出现的打字机文本图形。在 20 世纪 60 年代，计算机有了较弱的图形处理硬件，ASCII 被用于表示图形。今天，ASCII 文本图形继续作为因特网上的一种表现形式，你可以在网上找到各种创意的例子。

这个项目用 Python 创建一个程序，从图像生成 ASCII 文本图形。该程序让你指定输出（文本列数）的宽度，并设置垂直比例因子。它也支持两种灰度值到 ASCII 字符的映射：稀疏的 10 级映射和更精细校正的 70 级映射。

要从图像生成 ASCII 文本图形，需要学习如何做到以下几点：

- 用 Pillow 将彩色图像转换成灰度图像，它是 Python 的图像库（PIL）的一个分支；
- 使用 numpy 计算灰度图像的平均亮度；
- 用一个字符串作为灰度值的快速查找表。

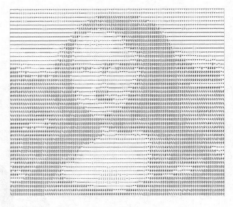

图 6-1 ASCII 文本图形的例子

6.1 工作原理

该项目利用了这样一个事实:从远处看,我们将灰度图像看成是它们亮度的平均值。例如,在图 6-2 中,可以看到一个建筑物的灰度图像,在它旁边,是填充了该建筑物图像的平均亮度值的图像。如果穿过一间屋子来看这两幅图像,它们看起来相似。

ASCII 文本图形的生成方法是,将图像分割成小块,并用 ASCII 字符替换一小块的平均 RGB 值。从远处看,因为眼睛的分辨率有限,我们大致会丢失细节,看到 ASCII 文本图形中的"平均"值,否则文本图形看起来就不那么真实。

图 6-2 灰度图像的平均值

该程序将给定的图像先转换为 8 位的灰度，让每个像素有一个灰度值，范围在 [0,255]（8 位整数的范围）。将这个 8 位值看成是亮度，0 表示黑色，255 表示白色，中间值是不同程度的灰色。

接着，将该图像分割成 M×N 个小块构成的网格（其中，M 和 N 是 ASCII 文本图形中的行和列编号）。然后程序计算网格中每个小块的平均亮度值，通过预定义的一些有梯度的 ASCII 字符（一组不断增加的值）来表示[0,255]范围的灰度值，与适当的 ASCII 字符匹配。它将用这些值作为亮度值的查找表。

完成的 ASCII 文本图形只是一些文本行。要显示文本，就要用到 Courier 这样的等宽字体，因为如果每个文本字符宽度不相同，图像中字符将无法正确地按网格排列，会得到间隔不均和失真的输出。

所用字体的"横纵比"（宽度与高度之比）也会影响最终图像。如果一个字符所占空间的横纵比与该字符取代的图像小块的横纵比不同，则最终的 ASCII 字符图形会出现失真。实际上，你试图用一个 ASCII 字符来替换图像小块，所以它们的形状要匹配。例如，如果将图像分割成正方形小块，然后用一种高度拉伸的字体替换每个小块，最终的结果将出现垂直拉伸。

为了解决这个问题，需要缩放网格中的行数，以匹配 Courier 的长宽比。（可以向程序发送命令行参数，修改缩放，以匹配其他字体。）

总之，下面是程序生成 ASCII 文本图形的步骤：
1. 将输入图像转成灰度；
2. 将图像分成 M×N 个小块；
3. 修正 M（行数），以匹配图像和字体的横纵比；
4. 计算每个小块图像的平均亮度，然后为每个小块查找合适的 ASCII 字符；
5. 汇集各行 ASCII 字符串，将它们打印到文件，形成最终图像。

6.2 所需模块

这个项目将使用 Pillow（Python 图像库的友好分支）来读取图像，访问它们的底层数据，创建并修改它们。还将使用 numpy 库来计算平均值。

6.3 代码

开始先定义灰度等级，用于生成 ASCII 文本图形。然后，考虑如何将图像分割成小块，以及如何计算这些小块的平均亮度。接下来，用 ASCII 字符替换小块，生成最终的输出。最后，为程序设置命令行解析，允许用户指定输出尺寸、输出文件名，等等。

整个项目的代码，请参阅 6.4 节。

6.3.1 定义灰度等级和网格

创建程序的第一步是，先定义两种灰度等级作为全局值，用于将亮度值转换为 ASCII 字符。

```
# 70 levels of gray
```
❶ `gscale1 = "$@B%8&WM#*oahkbdpqwmZO0QLCJUYXzcvunxrjft/\|()1{}[]?-_+~<>i!lI;:,\"^ `". "`

```
# 10 levels of gray
```
❷ `gscale2 = "@%#*+=-:. "`

❶行的值 gscale1 是 70 级的灰度梯度，❷行的 gscale2 是简单的 10 级灰度梯度。这两个值保存为字符串，包含一组字符串，从最黑暗变到最亮（要了解如何用字符表示灰度值的更多内容，参见 Paul Bourke 的《Character Representation of Grey Scale Images》，网址在 http://paulbourke.net/dataformats/asciiart/）。

既然有了灰度梯度，就可以准备图像。下面的代码打开图像，并分割成网格：

```
    # open the image and convert to grayscale
❶   image = Image.open(fileName).convert("L")
    # store the image dimensions
❷   W, H = image.size[0], image.size[1]
    # compute the tile width
❸   w = W/cols
    # compute the tile height based on the aspect ratio and scale of the font
❹   h = w/scale
    # compute the number of rows to use in the final grid
❺   rows = int(H/h)
```

在❶行，Image.open()打开输入图像文件，Image.convert()将该图像转换为灰度图像。"L" 代表 luminance，是图像亮度的单位。

在❷行，保存输入图像的宽度和高度。在❸行，根据用户指定的列数（cols），计算小块的宽度（如果用户没有在命令行中设置其他值，程序默认使用 80 列）。对于❸行的除法，使用浮点而不是整数除法，避免在计算小块尺寸时的截断误差。

知道小块的宽度后，在❹行利用垂直比例系数（作为 scale 传入），计算它的高度。在❺行，用这个网格高度来计算行数。

比例系数确定每个小块的大小，以匹配用于显示文本的字体的横纵比，这样最终图像不会失真。scale 的值可以作为参数传入，或者设置为默认值 0.43，在用 Courier 显示结果时，效果很好。

6.3.2 计算平均亮度

接下来，计算灰度图像中每一小块的平均亮度。函数 getAverageL()完成这项工作。

```
❶   def getAverageL(image):
        # get the image as a numpy array
❷       im = np.array(image)
        # get the dimensions
❸       w,h = im.shape
        # get the average
❹       return np.average(im.reshape(w*h))
```

在❶行，图像小块作为 PIL Image 对象传入。在❷行，将 image 转换成一个 numpy 数组，此时 im 成为一个二维数组，包含每个像素的亮度。在❸行，保存该图像的尺寸（宽度和高度）。在❹行，numpy.average()计算该图像中的亮度平均值，做法是用 numpy.reshape()先将维度为宽和高(w，h)的二维数组转换成扁平的一维，其长度是宽度乘以高度(w*h)。然后 numpy.average()调用对这些数组值求和并计算平均值。

6.3.3 从图像生成 ASCII 内容

程序的主要部分负责从图像生成 ASCII 内容。

```
        # an ASCII image is a list of character strings
❶       aimg = []
        # generate the list of tile dimensions
        for j in range(rows):
❷           y1 = int(j*h)
            y2 = int((j+1)*h)
            # correct the last tile
            if j == rows-1:
                y2 = H
            # append an empty string
❸           aimg.append("")
            for i in range(cols):
                # crop the image to fit the tile
❹               x1 = int(i*w)
                x2 = int((i+1)*w)
                # correct the last tile
❺               if i == cols-1:
                    x2 = W
                # crop the image to extract the tile into another Image object
❻               img = image.crop((x1, y1, x2, y2))
                # get the average luminance
❼               avg = int(getAverageL(img))
                # look up the ASCII character for grayscale value (avg)
                if moreLevels:
❽                   gsval = gscale1[int((avg*69)/255)]
                else:
❾                   gsval = gscale2[int((avg*9)/255)]
                # append the ASCII character to the string
❿               aimg[j] += gsval
```

在程序的这一部分，ASCII 图像先作为一个字符串列表保存，该列表在❶行初始化。接下来，按计算好的图像小块行数迭代遍历，在❷行和随后一行中，计算每个图像小块的起始和结束 y 坐标。虽然这些是浮点运算，但在传给图像裁剪方法之前，将它们截断为整数。

接着，因为只有当图像的宽度是列数的整数倍时，图像分割成小块时，边缘的小块才有相同的大小，所以在最后一行校正小块的 y 坐标，将 y 坐标设置为图像的实际高度。这样做确保了图像顶部的边缘不被截断。

在❸行，为 ASCII 图像添加一个空字符串，作为一种紧凑的方式来表示图像的当前行。接下来会填充这个字符串（将字符串作为字符的列表）。

在❹行和下一行，计算每个小块的左、右 x 坐标，在❺行，为最后一小块校正 x 坐标，原因和校正 y 坐标时一样。在❻行，用 image.crop()提取图像小块，然后将该小块传入 getAverageL()函数❼，该函数在 6.3.2 节中定义，取得小块的平均亮度。在❾行，将平均亮度值从[0，255]缩小至[0，9]（默认 10 级灰度梯度值的范围）。然后，用 gscale2（保存的梯度字符串）作为查找表，找到对应的 ASCII 值。❽行类似，不同之处在于，只有命令行标志设置为使用 70 级梯度时，才会用它。最后，在❿行，在文本行中添加找到的 ASCII 值 gsval，代码循环，直到处理完所有行。

6.3.4　命令行选项

接下来，为程序定义一些命令行选项。这段代码使用内置的 argparse 类：

```
        parser = argparse.ArgumentParser(description="descStr")
        # add expected arguments
❶       parser.add_argument('--file', dest='imgFile', required=True)
❷       parser.add_argument('--scale', dest='scale', required=False)
❸       parser.add_argument('--out', dest='outFile', required=False)
❹       parser.add_argument('--cols', dest='cols', required=False)
❺       parser.add_argument('--morelevels', dest='moreLevels', action='store_true')
```

在❶行，包含指定图像文件输入的选项（唯一必须的参数），❷行设置垂直比例因子，❸行设置输出文件名，❹行设置 ASCII 输出中的文本列数。在❺行，添加 --morelevels 选项，让用户选择更多层次的灰度梯度。

6.3.5　将 ASCII 文本图形字符串写入文本文件

最后，将生成的 ASCII 字符串列表，写入一个文本文件：

```
        # open a new text file
❶       f = open(outFile, 'w')
        # write each string in the list to the new file
❷       for row in aimg:
            f.write(row + '\n')
        # clean up
❸       f.close()
```

在❶行，使用内置的 open()方法，打开一个新的文本文件用于写入。然后在❷行，迭代遍历列表中的每个字符串，将它写入文件，在❸行，关闭文件对象，释放系统资源。

6.4 完整代码

下面是完整的 ASCII 文本图形程序。也可以从 https://github.com/electronut/pp/blob/master/ascii/ascii.py 下载该项目的代码。

```python
import sys, random, argparse
import numpy as np
import math
from PIL import Image

# grayscale level values from:
# http://paulbourke.net/dataformats/asciiart/

# 70 levels of gray
gscale1 = "$@B%8&WM#*oahkbdpqwmZO0QLCJUYXzcvunxrjft/\|()1{}[]?-_+~<>i!lI;:,\"^`'. "
# 10 levels of gray
gscale2 = '@%#*+=-:. '

def getAverageL(image):
    """
    Given PIL Image, return average value of grayscale value
    """
    # get image as numpy array
    im = np.array(image)
    # get the dimensions
    w,h = im.shape
    # get the average
    return np.average(im.reshape(w*h))

def covertImageToAscii(fileName, cols, scale, moreLevels):
    """
    Given Image and dimensions (rows, cols), returns an m*n list of Images
    """
    # declare globals
    global gscale1, gscale2
    # open image and convert to grayscale
    image = Image.open(fileName).convert('L')
    # store the image dimensions
    W, H = image.size[0], image.size[1]
    print("input image dims: %d x %d" % (W, H))
    # compute tile width
    w = W/cols
    # compute tile height based on the aspect ratio and scale of the font
    h = w/scale
    # compute number of rows to use in the final grid
    rows = int(H/h)

    print("cols: %d, rows: %d" % (cols, rows))
    print("tile dims: %d x %d" % (w, h))

    # check if image size is too small
    if cols > W or rows > H:
        print("Image too small for specified cols!")
        exit(0)
```

```python
        # an ASCII image is a list of character strings
        aimg = []
        # generate the list of tile dimensions
        for j in range(rows):
            y1 = int(j*h)
            y2 = int((j+1)*h)
            # correct the last tile
            if j == rows-1:
                y2 = H
            # append an empty string
            aimg.append("")
            for i in range(cols):
                # crop the image to fit the tile
                x1 = int(i*w)
                x2 = int((i+1)*w)
                # correct the last tile
                if i == cols-1:
                    x2 = W
                # crop the image to extract the tile into another Image object
                img = image.crop((x1, y1, x2, y2))
                # get the average luminance
                avg = int(getAverageL(img))
                # look up the ASCII character for grayscale value (avg)
                if moreLevels:
                    gsval = gscale1[int((avg*69)/255)]
                else:
                    gsval = gscale2[int((avg*9)/255)]
                # append the ASCII character to the string
                aimg[j] += gsval

    # return text image
    return aimg

# main() function
def main():
    # create parser
    descStr = "This program converts an image into ASCII art."
    parser = argparse.ArgumentParser(description=descStr)
    # add expected arguments
    parser.add_argument('--file', dest='imgFile', required=True)
    parser.add_argument('--scale', dest='scale', required=False)
    parser.add_argument('--out', dest='outFile', required=False)
    parser.add_argument('--cols', dest='cols', required=False)
    parser.add_argument('--morelevels', dest='moreLevels', action='store_true')

    # parse arguments
    args = parser.parse_args()

    imgFile = args.imgFile
    # set output file
    outFile = 'out.txt'
    if args.outFile:
        outFile = args.outFile
    # set scale default as 0.43, which suits a Courier font
    scale = 0.43
    if args.scale:
        scale = float(args.scale)
    # set cols
    cols = 80
    if args.cols:
```

```
        cols = int(args.cols)
    print('generating ASCII art...')
    # convert image to ASCII text
    aimg = covertImageToAscii(imgFile, cols, scale, args.moreLevels)

    # open a new text file
    f = open(outFile, 'w')
    # write each string in the list to the new file
    for row in aimg:
        f.write(row + '\n')
    # clean up
    f.close()
    print("ASCII art written to %s" % outFile)

# call main
if __name__ == '__main__':
    main()
```

6.5 运行 ASCII 文本图形生成程序

要运行编写好的程序，输入类似下面这样的命令，将 data/robot.jpg 替换为你想使用的图像文件的相对路径。

```
$ python ascii.py --file data/robot.jpg --cols 100
```

图 6-3 展示了 ASCII 文本图形，它是左侧的 robot.jpg 的结果。

图 6-3　ascii.py 的运行示例

现在，你可以创建自己的 ASCII 文本图形了！

6.6 小结

在这个项目中，我们学习了如何从任意的输入图像生成 ASCII 文本图形。我们还学习了如何计算平均亮度值，将图像转换成灰度，以及如何基于灰度值用字符替换一小块图像。创建自己的 ASCII 文本图形，祝你玩得开心！

6.7 实验

这里有一些想法，可以进一步探索 ASCII 文本图形。

1. 用命令行选项--scale1.0 运行该程序。生成的图像看起来如何？实验不同的 scale 值。将输出复制到一个文本编辑器，尝试设置不同的（固定宽度）字体，看看这样做如何影响最终图形的外观。

2. 为程序添加命令行选项--invert，反转 ASCII 文本图形的输入值，使黑色变成白色，反之亦然（提示：在查找时用 255 减去小块的亮度值）。

3. 在这个项目中，基于两个字符硬编码的梯度创建灰度值查找表。实现一个命令行选项，用不同的字符梯度来创建 ASCII 文本图形，就像这样：

```
python3 ascii.py --map "@$%^`."
```

前面这样的梯度用给定的 6 个字符梯度，创建了一个 ASCII 输出，其中"@"映射到亮度值 0，"."映射到亮度值 255。

第 7 章
照片马赛克

我在六年级时,看到一张类似图 7-1 的图像,但不太明白它是什么。眯着眼睛看了一阵后,我终于想通了(把书倒过来,离远一点看它)。

照片马赛克是一张图像,它被分割成长方形的网格,每个长方形由另一张匹配"目标"的图像(最终希望出现在照片马赛克中的图像)替代。换言之,如果从远处看照片马赛克,会看到目标图像;但如果走近,会看到该图像实际上包含许多较小的图像。

这个迷题有解是因为人眼的工作方式。图 7-1 中的低分辨率块状图像,靠近了很难识别,但如果从远处看,就知道它代表什么,因为看到的细节较少,就使得边缘越光滑。照片马赛克的原理是相似的。从远处看,图像看起来正常,但走近时,秘密揭开了:每"块"都是一个独特的图像!

本项目中,我们将学习如何用 Python 创建照片马赛克。我们将目标图像划分成较小图像的网格,并用适当的图像替换网格中的每一小块,创建原始图像的照片马赛克。你可以指定网格的尺寸,并选择输入图像是否可以在马赛克中重复使用。

在这个项目中,你将学习如何做到以下几点:

- 用 Python 图像库(PIL)创建图像;

- 计算图像的平均 RGB 值；
- 剪切图像；
- 通过粘贴另一张图像来替代原图像的一部分；
- 利用平均距离测量来比较 RGB 值。

图 7-1　令人费解的图像

7.1　工作原理

要创建照片马赛克，就从目标图像的块状低分辨率版本开始（因为在高分辨率的图像中，小块图像的数量会太大）。该图像的分辨率将决定马赛克的维度 M×N（M 是行数，N 是列数）。接着，根据这种方法替换原始图像中的每一小块：

1. 读入一些小块图像，它们将取代原始图像中的小块；
2. 读入目标图像，将它分割成 M×N 的小块网格；
3. 对于每个小块，从输入的小块图像中找到最佳匹配；
4. 将选择的输入图像安排在 M×N 的网格中，创建最终的照片马赛克。

7.1.1　分割目标图像

按照图 7-2 中的方案，开始将目标图像划分成 M×N 的网格。

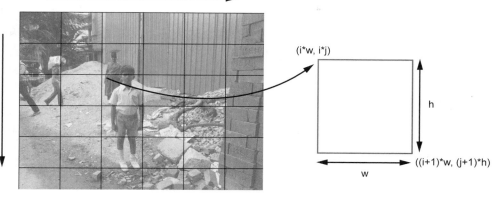

图 7-2 分割目标图像

图 7-2 中的图像展示了如何将原始图像分割成小块的网格。x 轴表示网格的列，y 轴表示网格的行。

现在，看看如何计算网格中一个小块的坐标。下标为（i，j）的小块，左上角坐标为 (i*w, i*j)，右下角坐标为((i+1)*w, (j+1)*h)，其中 w 和 h 分别是小块的宽度和高度。PIL 可以用这些数据，从原图像创建小块。

7.1.2 平均颜色值

图像中的每个像素都有颜色，由它的红、绿、蓝值来表示。在这个例子中，使用 8 位的图像，因此每个部分有 8 位值，范围在[0,255]。如果一幅图像共有 N 个像素，平均 RGB 计算如下：

$$r,g,b\left(\frac{r_1+r_2+\cdots+r_N}{N}, \frac{g_1+g_2+\cdots+g_N}{N}, \frac{b_1+b_2+\cdots+b_N}{N}\right)$$

请注意，平均 RGB 也是一个三元组，不是标量或一个数字，因为平均值是针对每个颜色成分分别计算的。计算平均 RGB 是为了匹配图像小块和目标图像。

7.1.3 匹配图像

对于目标图像中的每个小块，需要在用户指定的输入文件夹的一些图像中，找到一幅匹配的图像。要确定两个图像是否匹配，就使用平均 RGB 值。最接近的匹配就是最接近平均 RGB 值的图像。

要做到这一点，最简单的方法是计算一个像素中 RGB 值之间的距离，以便从输入图像中找到最佳匹配。对于几何中的三维点，可以用以下的距离计算方法：

$$D_{1,2} = \sqrt{(r_1-r_2)^2 + (g_1-g_2)^2 + (b_1-b_2)^2}$$

这里计算了点 (r_1, g_1, b_1) 和 (r_2, g_2, b_2) 之间的距离。给定一个目标图像的平均 RGB 值，以及来自输入图像的平均 RGB 值列表，你可以使用线性搜索和三维点的距离计算，来找到最匹配的图像。

7.2 所需模块

这个项目将使用 Pillow 读入图像，访问其底层数据，创建和修改图像。还会用 numpy 来操作图像数据。

7.3 代码

首先读入那些小块图像，它们将用于创建照片马赛克。接下来，计算图像的平均 RGB 值，然后将目标图像分割成图像网格，为小块找到最佳匹配。最后，组装图像小块，创建最终的照片马赛克。要查看完整的项目代码，请直接跳到 7.4 节。

7.3.1 读入小块图像

首先，从给定的文件夹中读入输入图像。下面是具体做法：

```
def getImages(imageDir):
    """
    given a directory of images, return a list of Images
    """
❶   files = os.listdir(imageDir)
    images = []
    for file in files:
❷       filePath = os.path.abspath(os.path.join(imageDir, file))
        try:
            # explicit load so we don't run into resource crunch
❸           fp = open(filePath, "rb")
            im = Image.open(fp)
            images.append(im)
            # force loading the image data from file
❹           im.load()
            # close the file
❺           fp.close()
        except:
            # skip
            print("Invalid image: %s" % (filePath,))
    return images
```

在❶行，用 os.listdir()将 imageDir 目录中的文件放入一个列表。接下来，迭代遍历列表中的每个文件，将它载入一个 PIL Image 对象。

在❷行，用 os.path.abspath()和 os.path.join()来获取图像的完整文件名。这个习惯用法在 Python 中经常使用，以确保代码既能在相对路径下工作（如\foo\bar），也能在绝对路径下工作（c:\foo\bar\），并且能跨操作系统，不同的操作系统有不同的目录命名惯例（Windows 用\而 Linux 用/）。

要将文件加载为 PIL 的 Image 对象，可以将每个文件名传入 Image.open()方法，但如果照片马赛克文件夹中有几百张甚至几千张图片，这样做非常消耗资源。作为替代，可以用 Python 打开每个小块图像，利用 Image.open()将文件句柄 fp 传入 PIL()。图像加载后，关闭文件句柄并释放系统资源。

在❸行，用 open()打开图像文件。在随后的几行中，将文件句柄传入 Image.open()，将得到的图像 im 存入一个数组。

因为 open()是一个惰性操作，所以在❹行调用 Image.load()，强制加载 im 中的图像数据。它确定了图像，但实际上没有读取全部图像数据，直到尝试使用该图像的时候才会那么做。

在❺行，关闭文件句柄，释放系统资源。

7.3.2 计算输入图像的平均颜色值

读入输入图像后，需要计算它们的平均颜色值，以及目标图像中的每个小块的值。创建一个方法 getAverageRGB()来计算这两个值。

```
def getAverageRGB(image):
    """
    return the average color value as (r, g, b) for each input image
    """
    # get each tile image as a numpy array
❶   im = np.array(image)
    # get the shape of each input image
❷   w,h,d = im.shape
    # get the average RGB value
❸   return tuple(np.average(im.reshape(w*h, d), axis=0))
```

在❶行，用 numpy 将每个 Image 对象转换为数据数组。返回的 numpy 数组形为(w, h, d)，其中，w 是图像的宽度，h 是高度，d 是深度，在这个例子中，是 RGB 图像的 3 个单位（分别对应 R, G 和 B）。在❷行保存 shape 元组，然后计算平均 RGB 值，将这个数组变形为更方便的形状(w*h, d)，这样就可以用 numpy.average()计算出使用平均值❸。

7.3.3 将目标图像分割成网格

现在，需要将目标图像分割成 M×N 网格，包含更小的图像。让我们创建一个方法来实现。

```
def splitImage(image, size):
    """
    given the image and dimensions (rows, cols), return an m*n list of images
    """
❶   W, H = image.size[0], image.size[1]
❷   m, n = size
❸   w, h = int(W/n), int(H/m)
    # image list
    imgs = []
    # generate a list of dimensions
    for j in range(m):
        for i in range(n):
            # append cropped image
❹           imgs.append(image.crop((i*w, j*h, (i+1)*w, (j+1)*h)))
    return imgs
```

首先，在❶行得到目标图像的维度，在❷行得到尺寸。在❸行，用基本除法计算目标图像中每一小块的尺寸。

现在，需要迭代遍历网格的维度，分割并将每一小块保存为单独的图像。在❹行，image.crop()利用左上角图像坐标和裁剪图像的维度作为参数，剪裁出图像的一部分（在 7.1.1 小节中讨论）。

7.3.4 寻找小块的最佳匹配

现在，让我们从输入图像的文件夹中，找到小块的最佳匹配。创建一个工具方法 getBestMatchIndex()，如下所示：

```
def getBestMatchIndex(input_avg, avgs):
    """
    return index of the best image match based on average RGB value distance
    """

    # input image average
    avg = input_avg

    # get the closest RGB value to input, based on RGB distance
    index = 0
❶   min_index = 0
❷   min_dist = float("inf")
❸   for val in avgs:
❹       dist = ((val[0] - avg[0])*(val[0] - avg[0]) +
                (val[1] - avg[1])*(val[1] - avg[1]) +
                (val[2] - avg[2])*(val[2] - avg[2]))
❺       if dist < min_dist:
            min_dist = dist
            min_index = index
        index += 1

    return min_index
```

需要从列表 avgs 中，找到最匹配平均 RGB 值 input_avg 的。avgs 是小块图像平均 RGB 值的列表。

为了找到最佳匹配，比较这些输入图像的平均 RGB 值。在❶行和❷行，将最接近的匹配下标初始化为 0，最小距离初始化为无穷大。该测试在第一次总是会通过，因为任何距离都小于无穷大。在❸行，遍历平均值列表中的值，并开始在❹行用标准公式计算距离（比较距离的平方，以减少计算时间）。在❺行，如果计算的距离小于保存的最小距离 min_dist，它就被替换为新的最小距离。迭代结束时，就得到了平均 RGB 值列表 args 中，最接近 input_avg 的下标。现在可以利用这个下标，从小块图像的列表中选择匹配的小块图像了。

7.3.5　创建图像网格

在继续创建照片马赛克之前，还需要一个工具方法。createImageGrid()方法将创建大小为 M×N 的图像网格。这个图像网格是最终的照片马赛克图像，利用选择的小块图像列表来创建。

```
def createImageGrid(images, dims):
    """
    given a list of images and a grid size (m, n), create a grid of images
    """
❶   m, n = dims

    # sanity check
    assert m*n == len(images)

    # get the maximum height and width of the images
    # don't assume they're all equal
❷   width = max([img.size[0] for img in images])
    height = max([img.size[1] for img in images])

    # create the target image
❸   grid_img = Image.new('RGB', (n*width, m*height))

    # paste the tile images into the image grid
    for index in range(len(images)):
❹       row = int(index/n)
❺       col = index - n*row
❻       grid_img.paste(images[index], (col*width, row*height))

    return grid_img
```

在❶行，取得网格的尺寸，然后用 assert 检查，提供给 createImageGrid()的图像数量是否符合网格的大小（assert 方法检查代码中的假定，特别是在开发和测试过程中的假定）。现在你有一个小块图像列表，基于最接近的 RGB 值，你将用它来创建一幅图像，表现照片马赛克。由于大小差异，某些选定的图像可能不会正好填充一个小块，但这不会是一个问题，因为你首先用黑色背景填充小块。

在❷行和下面一行,计算小块图像的最大宽度和高度(你没有对选择的输入图像的大小做出任何假定,无论它们相同或不同,代码都能工作),如果输入图像不能完全填充小块,小块之间的空间将显示为背景色,默认是黑色。

在❸行,创建一个空的 Image,大小符合网格中的所有图像。小块图像会粘贴到这个图像。然后填充图像网格。在❻行,循环遍历选定的图像,利用 Image.paste() 方法,将它们粘贴到相应的网格中。Image.paste() 的第一个参数是要粘贴的 Image 对象,第二个参数是左上角的坐标。现在,你需要搞清楚小块图像要粘贴到图像网格的行和列。为了做到这一点,将图像下标表示为行和列。小块在图像网格中的下标由 N*row + col 给出,其中 N 是一行的小块数,(row, col)是在该网格中的坐标。在❹行,利用前面的公式给出行,❺行给出了列。

7.3.6 创建照片马赛克

现在,有了所有必需的工具方法,让我们编写一个 main 函数,创建照片马赛克。

```python
def createPhotomosaic(target_image, input_images, grid_size, reuse_images=True):
    """
    creates a photomosaic given target and input images
    """

    print('splitting input image...')
    # split the target image into tiles
❶   target_images = splitImage(target_image, grid_size)

    print('finding image matches...')
    # for each tile, pick one matching input image
    output_images = []
    # for user feedback
    count = 0
❷   batch_size = int(len(target_images)/10)
    # calculate the average of the input image
    avgs = []
    for img in input_images:
❸       avgs.append(getAverageRGB(img))

    for img in target_images:
        # compute the average RGB value of the image
❹       avg = getAverageRGB(img)
        # find the matching index of closest RGB value
        # from a list of average RGB values
❺       match_index = getBestMatchIndex(avg, avgs)
❻       output_images.append(input_images[match_index])
        # user feedback
❼       if count > 0 and batch_size > 10 and count % batch_size is 0:
            print('processed %d of %d...' %(count, len(target_images)))
        count += 1
        # remove the selected image from input if flag set
❽       if not reuse_images:
            input_images.remove(match)
```

```
    print('creating mosaic...')
    # create photomosaic image from tiles
❾  mosaic_image = createImageGrid(output_images, grid_size)

    # display the mosaic
    return mosaic_image
```

createPhotomosaic()方法的输入是目标图像、输入图像列表、生成照片马赛克的大小，以及一个标志，该标志表明图像是否可以复用。在❶行，它将目标图像分割成一个网格。图像被分割后，针对每个小块，从输入文件夹中寻找匹配的图像（因为这个过程可能很长，所以提供反馈给用户，让他们知道程序仍在工作）。

在❷行，将 batch_size 设置为小块图像总数的十分之一。在❼行，该变量将用于向用户更新信息（选择十分之一是任意的，只是一种方式让程序说："我还活着。"每次程序处理了图像的十分之一，就打印一条消息，表明它仍在运行）。

在❸行，为输入文件夹中的每个图像计算平均 RGB 值，并保存在列表 avgs 中。然后，开始迭代遍历目标图像网格中的每个小块。对于每个小块，计算平均 RGB 值❹。然后，在❺行，在输入图像的平均值列表中，寻找该值的最佳匹配。返回的结果是一个下标，在❻行用它取得 Image 对象，并保存在列表中。

在❼行，每处理 batch_size 个图像，就为用户打印一条消息。在❽行，如果 reuse_images 标志设置为 False，就从列表中删除选定的输入图像，这样就不会在另一个小块中重用（如果有广泛的输入图像可选，这种方式效果最好）。最后，在❾行，组合图像创建最终的照片马赛克。

7.3.7 添加命令行选项

该程序的 main()方法支持这些命令行选项：

```
# parse arguments
parser = argparse.ArgumentParser(description='Creates a photomosaic from
                                input images')
# add arguments
parser.add_argument('--target-image', dest='target_image', required=True)
parser.add_argument('--input-folder', dest='input_folder', required=True)
parser.add_argument('--grid-size', nargs=2, dest='grid_size', required=True)
parser.add_argument('--output-file', dest='outfile', required=False)
```

这段代码包含 3 个必需的命令行参数：目标图像的名称、输入图像文件夹的名称，以及网格尺寸。第 4 个参数是可选的文件名。如果省略该文件名，照片马赛克将写入文件 mosaic.png。

7.3.8 控制照片马赛克的大小

要解决的最后一个问题是照片马赛克的大小。如果基于目标图像中匹配的小块，盲目地将输入图像粘贴在一起，就会得到一个巨大的照片马赛克，比目标图像大得多。为了避免这种情况,调整输入图像的大小,以匹配网格中每个小块的大小(这样做还有一个好处,可以加快平均 RGB 的计算,因为用了较小的图像)。main()方法也进行这样的处理:

```
        print('resizing images...')
        # for given grid size, compute the maximum width and height of tiles
❶       dims = (int(target_image.size[0]/grid_size[1]),
                int(target_image.size[1]/grid_size[0]))
        print("max tile dims: %s" % (dims,))
        # resize
        for img in input_images:
❷           img.thumbnail(dims)
```

在❶行,根据指定的网格大小,计算目标图像的维度。随后,在❷行,用 PIL Image.thumbnail()方法来调整图像,以适应网格的大小。

7.4 完整代码

项目的完整代码可以在 https://github.com/electronut/pp/tree/master/photomosaic/photomosaic.py 找到。

```python
import sys, os, random, argparse
from PIL import Image
import imghdr
import numpy as np

def getAverageRGB(image):
    """
    return the average color value as (r, g, b) for each input image
    """
    # get each tile image as a numpy array
    im = np.array(image)
    # get the shape of each input image
    w,h,d = im.shape
    # get the average RGB value
    return tuple(np.average(im.reshape(w*h, d), axis=0))

def splitImage(image, size):
    """
    given the image and dimensions (rows, cols), returns an m*n list of images
    """
    W, H = image.size[0], image.size[1]
    m, n = size
    w, h = int(W/n), int(H/m)
    # image list
    imgs = []
    # generate a list of dimensions
    for j in range(m):
        for i in range(n):
            # append cropped image
            imgs.append(image.crop((i*w, j*h, (i+1)*w, (j+1)*h)))
    return imgs

def getImages(imageDir):
```

```python
    """
    given a directory of images, return a list of Images
    """
    files = os.listdir(imageDir)
    images = []
    for file in files:
        filePath = os.path.abspath(os.path.join(imageDir, file))
        try:
            # explicit load so we don't run into a resource crunch
            fp = open(filePath, "rb")
            im = Image.open(fp)
            images.append(im)
            # force loading image data from file
            im.load()
            # close the file
            fp.close()
        except:
            # skip
            print("Invalid image: %s" % (filePath,))
    return images

def getImageFilenames(imageDir):
    """
    given a directory of images, return a list of image filenames
    """
    files = os.listdir(imageDir)
    filenames = []
    for file in files:
        filePath = os.path.abspath(os.path.join(imageDir, file))
        try:
            imgType = imghdr.what(filePath)
            if imgType:
                filenames.append(filePath)
        except:
            # skip
            print("Invalid image: %s" % (filePath,))
    return filenames

def getBestMatchIndex(input_avg, avgs):
    """
    return index of the best image match based on average RGB value distance
    """

    # input image average
    avg = input_avg
    # get the closest RGB value to input, based on RGB distance
    index = 0
    min_index = 0
    min_dist = float("inf")
    for val in avgs:
        dist = ((val[0] - avg[0])*(val[0] - avg[0]) +
                (val[1] - avg[1])*(val[1] - avg[1]) +
                (val[2] - avg[2])*(val[2] - avg[2]))
        if dist < min_dist:
            min_dist = dist
            min_index = index
        index += 1

    return min_index
```

```python
def createImageGrid(images, dims):
    """
    given a list of images and a grid size (m, n), create a grid of images
    """
    m, n = dims

    # sanity check
    assert m*n == len(images)

    # get the maximum height and width of the images
    # don't assume they're all equal
    width = max([img.size[0] for img in images])
    height = max([img.size[1] for img in images])

    # create the target image
    grid_img = Image.new('RGB', (n*width, m*height))

    # paste the tile images into the image grid
    for index in range(len(images)):
        row = int(index/n)
        col = index - n*row
        grid_img.paste(images[index], (col*width, row*height))

    return grid_img


def createPhotomosaic(target_image, input_images, grid_size, reuse_images=True):
    """
    creates photomosaic given target and input images
    """

    print('splitting input image...')
    # split the target image into tiles
    target_images = splitImage(target_image, grid_size)

    print('finding image matches...')
    # for each tile, pick one matching input image
    output_images = []
    # for user feedback
    count = 0
    batch_size = int(len(target_images)/10)

    # calculate the average of the input image
    avgs = []
    for img in input_images:
        avgs.append(getAverageRGB(img))

    for img in target_images:
        # compute the average RGB value of the image
        avg = getAverageRGB(img)
        # find the matching index of closest RGB value
        # from a list of average RGB values
        match_index = getBestMatchIndex(avg, avgs)
        output_images.append(input_images[match_index])
        # user feedback
        if count > 0 and batch_size > 10 and count % batch_size is 0:
            print('processed %d of %d...' %(count, len(target_images)))
```

```python
            count += 1
            # remove the selected image from input if flag set
            if not reuse_images:
                input_images.remove(match)

    print('creating mosaic...')
    # create photomosaic image from tiles
    mosaic_image = createImageGrid(output_images, grid_size)

    # display the mosaic
    return mosaic_image

# gather our code in a main() function
def main():
    # command line arguments are in sys.argv[1], sys.argv[2], ...
    # sys.argv[0] is the script name itself and can be ignored
    # parse arguments
    parser = argparse.ArgumentParser(description='Creates a photomosaic from
                                    input images')
    # add arguments
    parser.add_argument('--target-image', dest='target_image', required=True)
    parser.add_argument('--input-folder', dest='input_folder', required=True)
    parser.add_argument('--grid-size', nargs=2, dest='grid_size', required=True)
    parser.add_argument('--output-file', dest='outfile', required=False)

    args = parser.parse_args()

    ###### INPUTS ######

    # target image
    target_image = Image.open(args.target_image)

    # input images
    print('reading input folder...')
    input_images = getImages(args.input_folder)

    # check if any valid input images found
    if input_images == []:
        print('No input images found in %s. Exiting.' % (args.input_folder, ))
        exit()

    # shuffle list to get a more varied output?
    random.shuffle(input_images)

    # size of the grid
    grid_size = (int(args.grid_size[0]), int(args.grid_size[1]))

    # output
    output_filename = 'mosaic.png'
    if args.outfile:
        output_filename = args.outfile

    # reuse any image in input
    reuse_images = True

    # resize the input to fit the original image size?
    resize_input = True

    ##### END INPUTS #####
```

```python
    print('starting photomosaic creation...')

    # if images can't be reused, ensure m*n <= num_of_images
    if not reuse_images:
        if grid_size[0]*grid_size[1] > len(input_images):
            print('grid size less than number of images')
            exit()
    # resizing input
    if resize_input:
        print('resizing images...')
        # for given grid size, compute the maximum width and height of tiles
        dims = (int(target_image.size[0]/grid_size[1]),
                int(target_image.size[1]/grid_size[0]))
        print("max tile dims: %s" % (dims,))
        # resize
        for img in input_images:
            img.thumbnail(dims)

    # create photomosaic
    mosaic_image = createPhotomosaic(target_image, input_images, grid_size,
                                     reuse_images)

    # write out mosaic
    mosaic_image.save(output_filename, 'PNG')

    print("saved output to %s" % (output_filename,))
    print('done.')

# standard boilerplate to call the main() function
# to begin the program
if __name__ == '__main__':
    main()
```

7.5 运行照片马赛克生成程序

下面是该程序的运行示例：

```
$ python photomosaic.py --target-image test-data/cherai.jpg --input-folder
test-data/set6/ --grid-size 128 128
reading input folder...
starting photomosaic creation...
resizing images...
max tile dims: (23, 15)
splitting input image...
finding image matches...
processed 1638 of 16384 ...
processed 3276 of 16384 ...
processed 4914 of 16384 ...
creating mosaic...
saved output to mosaic.png
done.
```

图 7-3（a）展示了目标图像，（b）展示了照片马赛克。在（c），可以看到照片马赛克的特写。

 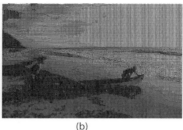 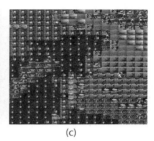

(a)   (b)   (c)

图 7-3   照片马赛克运行示例

7.6   小结

在这个项目中，我们学习了如何利用给定的目标图像和一组输入图像，创建一个照片马赛克。从远处看，照片马赛克像原来的图像，但走近了看，可以看到各个图像组成的马赛克。

7.7   实验

这里有一些方式，可以进一步探索照片马赛克。

1. 编写一个程序，创建图像的块状版本，类似图 7-1。

2. 利用本章的代码，通过粘贴匹配的图像创建照片马赛克，小块图像之间没有间隙。更艺术的表现形式，是在每个小块图像之间留出几个像素的均匀间隙。如何创建这样的间隙（提示：在计算最终的图像尺寸以及在 createImageGrid() 中粘贴时，考虑间隙的因素）？

3. 程序的大部分时间，用于从输入文件夹中寻找小块图像的最佳匹配。为了加快程序，getBestMatchIndex() 需要运行得更快。这个方法是对平均值（看成三维的点）列表进行简单的线性搜索。这个任务的一般问题就是最近邻居搜索。找到最近点有一种特别有效的方法，即 K-D 树搜索。SciPy 库有一个方便的类 scipy.spatial.KDTree，可以创建 K-D 并向它查询最近点的匹配。请尝试用 SciPy 的 K-D 树替代线性搜索（参见 http://docs.scipy.org/doc/scipy/reference/generated/scipy.spatial.KDTree.html）。

第 8 章
三维立体画

盯着图 8-1 看一分钟。除了随机的点,你还看到别的什么吗?图 8-1 是一张三维立体画,是制造三维假象的二维图像。 三维立体画通常包含一些重复的图案,仔细观察会被大脑解释为三维。如果你看不到任何图像效果,别担心。我也花了一段时间,做了一些尝试,才能看到(如果你在本书的印刷版本上看不到,请在这里尝试彩色的版本: https://github.com/electronut/pp/images/。标题的脚注提示了你应该看到的图像)。

本项目将用 Python 创建一张三维立体画。下面是本项目涉及的一些概念:

- 线性间距和深度知觉;
- 深度图;
- 用 Pillow 创建和编辑图像;
- 用 Pillow 绘制图像。

本项目生成的三维立体画设计为用 "墙眼" 方式观看。看到它们的最好方法,就是让眼睛聚焦在图像后面的点(如墙上)。有点神奇,一旦在这些图案中感知到某样东西,眼睛就会自动将它作为关注的焦点,如果三维图像已 "锁定",你很难对它视而不见的(如果你仍然无法看到图像,请看 Gene Levin 的文章 "How to View

Stereograms and Viewing Practice"[1]，或许有帮助）。

图 8-1　一张令人费解的图像，可能让你感到痛苦[2]

8.1　工作原理

三维立体画的工作原理是改变图像中图案之间的线性间距，从而产生深度的错觉。在观看三维立体画中的重复图案时，大脑会将间距解释为深度信息，如果有多个图案和不同的间距，尤其会这样。

8.1.1　感知三维立体画中的深度

如果你的眼睛汇聚在图像背后一个假想的点，大脑将左眼看到的一些点与右眼看到的另一些点匹配起来，你将会看到这些点位于图像之后的一个平面上。到该平面的感知距离取决于图案中的间距的数量。例如，图 8-2 展示了 3 行 A。这些 A 每行间的距离相等，但它们的水平间距从上至下增加。

如果用"墙眼"的方式来看，图 8-2 中最上面一行应该出现在纸后面，中间行应该看起来像在第一行后面一点，底部一行应该出现在最远的位置。文本"floating text"应该看起来"浮在"这几行顶部。

为什么大脑将这些图案的间距解读为深度？通常情况下，如果看远处的物体，你的双眼协作，聚焦并汇聚在同一点，双眼向内转，直接指向目标点。但用"墙眼"方式观看三维立体画时，聚焦和汇聚发生在不同的位置。眼睛专注于三维立体画，但大脑将重复的模式看成来自同一个虚拟（虚构的）对象，眼睛汇聚在图像背后的一个点，如图 8-3 所示。解耦的聚焦和汇聚叠加在一起，让你在三维立体画中看到深度。

---

[1] http://colorstereo.com/texts_.txt/practice.htm
[2] 隐藏图像是鲨鱼。

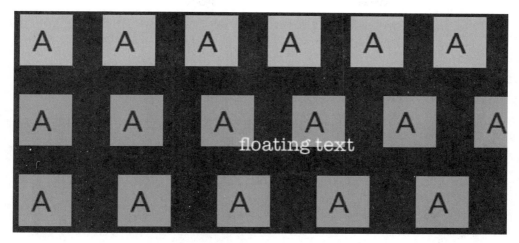

图 8-2　线性间距和深度知觉

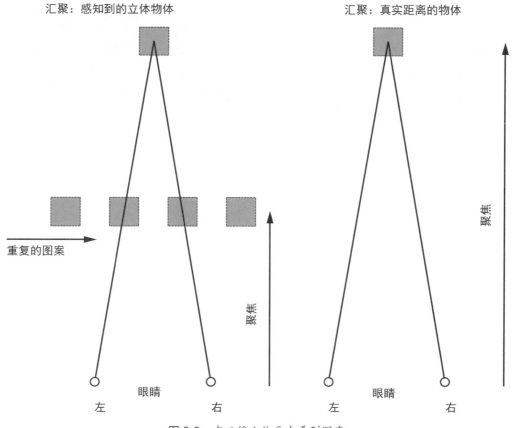

图 8-3　在三维立体画中看到深度

第 8 章　三维立体画

三维立体画的感知深度取决于像素的水平间距。因为图 8-2 中的第一行具有最近的间隔，它出现在其他行的前面。然而，如果点的间距在图像中是变化的，大脑将认为每个点处于不同的深度，所以我们会看到一个虚拟的三维图像。

8.1.2 深度图

"深度图"是这样一幅图像：其中每个像素的值表示深度值，即从眼睛到该像素表示的对象部分的距离。深度图往往表现为一幅灰度图，亮的区域表示近的点，暗的区域表示远的点，如图 8-4 所示。

图 8-4　深度图

注意，鲨鱼的鼻子是图像中最亮部分，似乎最接近你。朝向尾部的较暗区域看起来最远。

因为深度图表示从每个像素中心到眼睛的深度或距离，所以可以用它来获得与图像中像素位置相关联的深度值。我们知道，在图像中，水平偏移被认为是深度。所以，如果按照对应像素值深度值的比例，来偏移（图案）图像中的像素，就会对该像素产生与深度图一致的深度知觉。如果对所有像素这样做，最终就会将整个深度图编码到图像中，生成三维立体画。

深度图的每个像素存储了深度值，并且该值的分辨率取决于表示它的位数。因为本章采用常见的 8 位图像，深度值的范围是[0,255]。

顺便说一下，图 8-4 中的图像就是用于创建图 8-1 中的三维立体画的深度图。你很快就能学会自己如何做到这一点。

该项目的代码将遵循以下步骤：

1. 读入深度图；
2. 读入一幅平铺图像或创建一个"随机点"平铺图像；

3. 通过重复平铺图像创建一幅新图像。该图像的尺寸与深度图一致；
4. 对新图像中的每个像素，根据该像素相关联的深度值，将它按比例地向右移；
5. 将三维立体画写入一个文件。

8.2 所需模块

本项目使用 Pillow 读取图片，访问它们的底层数据，创建和修改图像。

8.3 代码

为了从输入的深度图生成三维立体画，首先重复一幅给定的平铺图像，生成一幅中间图像。接下来，生成一幅充满随机点的平铺图像。然后进入生成三维立体画的核心代码，即利用所提供的深度图中的信息，移动输入的图像。要查看完整的项目，请直接跳到 8.4 节。

8.3.1 重复给定的平铺图像

我们从利用 createTiledImage() 方法开始，通过平铺一个图形文件，创建一幅新的图像。图像尺寸由 dims 元组指定，该元组形式为（width, height）。

```
# tile a graphics file to create an intermediate image of a set size
def createTiledImage(tile, dims):
    # create the new image
❶   img = Image.new('RGB', dims)
    W, H = dims
    w, h = tile.size
    # calculate the number of tiles needed
❷   cols = int(W/w) + 1
❸   rows = int(H/h) + 1
    # paste the tiles into the image
    for i in range(rows):
        for j in range(cols):
❹           img.paste(tile, (j*w, i*h))
    # output the image
    return img
```

在❶行，利用提供的尺寸（dims）创建新的 Python 图像库（PIL）Image 对象。新图像的尺寸由元组 dims 给出，形式是（width, height）。接着，保存平铺图像和输出文件的宽度和高度。在❷行，确定列数，在❸行，确定中间图像所需的行数，方法是用最终图像的尺寸除以平铺图像的尺寸。除的结果每次加 1，如果输出图像的尺寸不是正好是平铺图像的整数倍，这也能确保右边最后的平铺图像不会缺失。如果没有这种预防措施，图像的右边可能被切断。然后，在❹行，循环遍历行和列，并用平铺图像填充它们。通过乘积(j*w, i*h)，确定平铺图像左上角的位置，这样它能对准行和列。完成后，该方法返回指定尺寸的 Image 对象，用输入图像 tile 平铺。

8.3.2 从随机圆创建平铺图像

如果用户不提供平铺图像，就利用 createRandomTile()方法，用随机圆圈创建一张平铺图像。

```
# create an image tile filled with random circles
def createRandomTile(dims):
    # create image
❶   img = Image.new('RGB', dims)
❷   draw = ImageDraw.Draw(img)
    # set the radius of a random circle to 1% of
    # width or height, whichever is smaller
❸   r = int(min(*dims)/100)
    # number of circles
❹   n = 1000
    # draw random circles
    for i in range(n):
        # -r makes sure that the circles stay inside and aren't cut off
        # at the edges of the image so that they'll look better when tiled
❺       x, y = random.randint(0, dims[0]-r), random.randint(0, dims[1]-r)
❻       fill = (random.randint(0, 255), random.randint(0, 255),
                random.randint(0, 255))
❼       draw.ellipse((x-r, y-r, x+r, y+r), fill)
    return img
```

在❶行，用 dim 给出的尺寸创建新的 Image 对象。用 ImageDraw.Draw() ❷在该图像中画圆圈，用宽或高中较小值的 1/100 作为半径，画圆圈❸（Python 的*运算符将 dim 元组中的宽度和高度值解包，这样就能传入到 min()方法中）。

在❹行，设置要画的圆圈数为 1000。然后调用 random.randint()，获得范围为[0, width-r]和[0, height-r]的两个随机整数，从而算出每个圆圈的 x 和 y 坐标❺。"-r"确保生成的圆圈保持在 width×height 的图像矩形内部。不带-r，画的圆圈可能就在图像边缘，这意味着它会被切掉一部分。如果平铺这样的图像来创建三维立体画，结果不会好看，因为两个平铺图像之间没有空间。

要生成一个随机圆圈，先画出轮廓，然后填充颜色。在❻行，在[0,255]的范围内随机选取 RGB 值，用选择颜色填充。最后，在❼行，用 draw 中的 ellipse()方法绘制每个圆圈。该方法的第一个参数是圆的边界矩形，它由左上角和右下角指定，分别为(x-r, y-r)和(x+r, y+r)，其中(x, y)是该圆的圆心，r 是半径。

让我们在 Python 解释器中测试这种方法。

```
>>> import autos
>>> img = autos.createRandomTile((256, 256))
>>> img.save('out.png')
>>> exit()
```

图 8-5 展示了测试的输出。

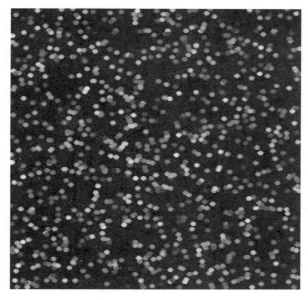

图 8-5　尝试运行 createRandomTile()

正如你在图 8-5 中看到的，我们已经创建了随机点的平铺图像。可以使用它来创建的三维立体画。

8.3.3　创建三维立体画

现在，让我们创建一些三维立体画。createAutostereogram()方法完成了大部分工作，如下所示：

```
def createAutostereogram(dmap, tile):
    # convert the depth map to a single channel if needed
❶   if dmap.mode is not 'L':
        dmap = dmap.convert('L')
    # if no image is specified for a tile, create a random circles tile
❷   if not tile:
        tile = createRandomTile((100, 100))
    # create an image by tiling
❸   img = createTiledImage(tile, dmap.size)
    # create a shifted image using depth map values
❹   sImg = img.copy()
    # get access to image pixels by loading the Image object first
❺   pixD = dmap.load()
    pixS = sImg.load()
    # shift pixels horizontally based on depth map
❻   cols, rows = sImg.size
    for j in range(rows):
        for i in range(cols):
```

```
❼                xshift = pixD[i, j]/10
❽                xpos = i - tile.size[0] + xshift
❾                if xpos > 0 and xpos < cols:
❿                    pixS[i, j] = pixS[xpos, j]
        # display the shifted image
        return sImg
```

在❶行，进行完整性检查，确保深度图和图像具有相同的尺寸。在❷行，如果用户没有提供平铺图像，就创建随机圆圈平铺图像。在❸行，创建一张平铺好的图像，符合提供的深度图的大小。然后，在❹行生成这张平铺好的图像的副本。

在❺行，调用 Image.load() 方法，将图像数据加载到内存中。该方法允许用形如 [i, j] 的二维数组来访问图像像素。在❻行，将图像的尺寸保存为行数和列数，将图像看成单个像素构成的网格。

三维立体画创建算法的核心在于，根据从深度图中收集的信息，移动平铺图像中像素的方式。要做到这一点，遍历平铺图像，处理每一个像素。在❼行，根据深度图 pixD 中的相关像素，查找偏移的值。然后将这个深度值除以 10，因为这里用的是 8 位深度图，这意味着深度的范围是 0 到 255。如果除以 10，得到的深度值范围是 0 到 25。由于深度图输入图像的尺寸通常是几百像素，所以这些偏移值很合适（尝试改变除数，看看它如何影响最终图像）。

在❽行，计算像素的新 x 位置，用平铺图像填充三维立体画。每隔 w 个像素，像素的值不断重复，由公式 $a_i = a_i + w$ 表示，其中的 a_i 是在 x 轴下标 *i* 处的给定像素的颜色（因为考虑的是像素行，而不是列，所以忽略 y 方向）。

要创建深度感，就要让间隔（或重复的间距）与该像素的深度图值成正比。这样在最终的三维立体画图像中，每个像素和它前一次（周期地）出现相比，偏移了 delta_i。这可以表示为 $b_i = b_{i-w+\delta_t}$

这里，b_i 表示最后的三维立体画图像中，下标 *i* 处给定像素的颜色值。这正是 ❽行所做的事。深度图值为 0（黑色）的像素没有偏移，被视为背景。

在❿行，用偏移的值替换每个像素。在❾行，检查确保没有试图访问不在图像中的像素，因为偏移，在图像边缘可能发生这种情况。

8.3.4　命令行选项

现在，我们来看看该程序的 main() 方法，其中提供了一些命令行选项。

```
    # create a parser
    parser = argparse.ArgumentParser(description="Autosterograms...")
    # add expected arguments
❶   parser.add_argument('--depth', dest='dmFile', required=True)
    parser.add_argument('--tile', dest='tileFile', required=False)
    parser.add_argument('--out', dest='outFile', required=False)
    # parse args
    args = parser.parse_args()
    # set the output file
    outFile = 'as.png'
```

```
    if args.outFile:
        outFile = args.outFile
    # set tile
    tileFile = False
    if args.tileFile:
        tileFile = Image.open(args.tileFile)
```

在❶行，像以前的项目一样，利用 argparse 为程序定义了一些命令行选项。一个必需的参数是深度图文件，两个可选的参数是平铺图像文件名和输出文件名。如果未指定平铺图像，程序会生成随机圆圈平铺图像。如果未指定输出文件名，则三维立体画会输出到 as.png 文件。

8.4 完整代码

下面是完整的三维立体画程序。也可以从 https://github.com/electronut/pp/blob/master/autos/autos.py 下载这段代码。

```python
import sys, random, argparse
from PIL import Image, ImageDraw

# create spacing/depth example
def createSpacingDepthExample():
    tiles = [Image.open('test/a.png'), Image.open('test/b.png'),
             Image.open('test/c.png')]
    img = Image.new('RGB', (600, 400), (0, 0, 0))
    spacing = [10, 20, 40]
    for j, tile in enumerate(tiles):
        for i in range(8):
            img.paste(tile, (10 + i*(100 + j*10), 10 + j*100))
    img.save('sdepth.png')

# create an image filled with random circles
def createRandomTile(dims):
    # create image
    img = Image.new('RGB', dims)
    draw = ImageDraw.Draw(img)
    # set the radius of a random circle to 1% of
    # width or height, whichever is smaller
    r = int(min(*dims)/100)
    # number of circles
    n = 1000
    # draw random circles
    for i in range(n):
        # -r makes sure that the circles stay inside and aren't cut off
        # at the edges of the image so that they'll look better when tiled
        x, y = random.randint(0, dims[0]-r), random.randint(0, dims[1]-r)
        fill = (random.randint(0, 255), random.randint(0, 255),
                random.randint(0, 255))
        draw.ellipse((x-r, y-r, x+r, y+r), fill)
    # return image
    return img

# tile a graphics file to create an intermediate image of a set size
```

```python
def createTiledImage(tile, dims):
    # create the new image
    img = Image.new('RGB', dims)
    W, H = dims
    w, h = tile.size
    # calculate the number of tiles needed
    cols = int(W/w) + 1
    rows = int(H/h) + 1
    # paste the tiles into the image
    for i in range(rows):
        for j in range(cols):
            img.paste(tile, (j*w, i*h))
    # output the image
    return img

# create a depth map for testing
def createDepthMap(dims):
    dmap = Image.new('L', dims)
    dmap.paste(10, (200, 25, 300, 125))
    dmap.paste(30, (200, 150, 300, 250))
    dmap.paste(20, (200, 275, 300, 375))
    return dmap

# given a depth map image and an input image,
# create a new image with pixels shifted according to depth
def createDepthShiftedImage(dmap, img):
    # size check
    assert dmap.size == img.size

# create shifted image
sImg = img.copy()
# get pixel access
pixD = dmap.load()
pixS = sImg.load()
# shift pixels output based on depth map
cols, rows = sImg.size
for j in range(rows):
    for i in range(cols):
        xshift = pixD[i, j]/10
        xpos = i - 140 + xshift
        if xpos > 0 and xpos < cols:
            pixS[i, j] = pixS[xpos, j]
    # return shifted image
    return sImg

# given a depth map (image) and an input image,
# create a new image with pixels shifted according to depth
def createAutostereogram(dmap, tile):
    # convert the depth map to a single channel if needed
    if dmap.mode is not 'L':
        dmap = dmap.convert('L')
    # if no image is specified for a tile, create a random circles tile
    if not tile:
        tile = createRandomTile((100, 100))
    # create an image by tiling
    img = createTiledImage(tile, dmap.size)
    # create a shifted image using depth map values
    sImg = img.copy()
```

```python
        # get access to image pixels by loading the Image object first
        pixD = dmap.load()
        pixS = sImg.load()
        # shift pixels horizontally based on depth map
        cols, rows = sImg.size
        for j in range(rows):
            for i in range(cols):
                xshift = pixD[i, j]/10
                xpos = i - tile.size[0] + xshift
                if xpos > 0 and xpos < cols:
                    pixS[i, j] = pixS[xpos, j]
        # return shifted image
        return sImg

# main() function
def main():
    # use sys.argv if needed
    print('creating autostereogram...')
    # create parser
    parser = argparse.ArgumentParser(description="Autosterograms...")
    # add expected arguments
    parser.add_argument('--depth', dest='dmFile', required=True)
    parser.add_argument('--tile', dest='tileFile', required=False)
    parser.add_argument('--out', dest='outFile', required=False)
    # parse args
    args = parser.parse_args()
    # set the output file
    outFile = 'as.png'
    if args.outFile:
        outFile = args.outFile
    # set tile
    tileFile = False
    if args.tileFile:
        tileFile = Image.open(args.tileFile)
    # open depth map
    dmImg = Image.open(args.dmFile)
    # create stereogram
    asImg = createAutostereogram(dmImg, tileFile)
    # write output
    asImg.save(outFile)

# call main
if __name__ == '__main__':
    main()
```

8.5 运行三维立体画生成程序

现在，我们用凳子（stool-depth.png）的深度图运行该程序。

```
$ python3 autos.py --depth data/stool-depth.png
```

图 8-6 左边展示了深度图，右边展示了生成的三维立体画。因为没有为平铺提供图像，这张三维立体画使用了随机生成的平铺图像。

图 8-6　autos.py 运行示例

现在，让我们给定一个平铺图像作为输入。像前面一样使用 stool-depth.png 深度图，但这一次，提供图像 escher-tile.jpg[①]作为平铺图像。

```
$ python3 autos.py --depth data/stool-depth.png --tile data/escher-tile.jpg
```

图 8-7 展示了输出。

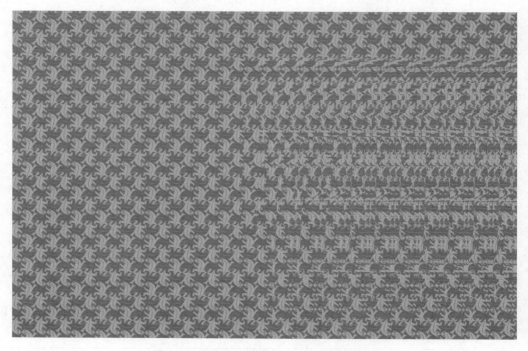

图 8-7　使用平铺图像的 autos.py 运行示例

[①] http://calculus-geometry.hubpages.com/hub/Free-M-C-Escher-Tessellation-Background-Patterns-Tiling-Lizard-Background/

8.6　小结

在本项目中，我们学习了如何创建三维立体画。给定深度图的图像，我们现在可以创建随机点的三维立体画，或用提供的图像来平铺。

8.7　实验

这里有一些方法，可以进一步探索三维立体画。

1．编程创建类似图 8-2 的图像，演示图像中线性间距的变化如何制造深度的幻觉（提示：利用图像平铺和 Image.paste()方法）。

2．为程序添加一个命令行选项，指定应用于深度图值的比例（回忆一下，代码中深度图值除以 10）。变更如何影响到三维立体画？

3．学习用 SketchUp 这样的工具，创建自己的三维模型深度图（http://sketchup.com/），或在线访问许多现成的 SketchUp 模型。利用 SketchUp 的 Fog 选项来创建你的深度图。如需帮助，看看这个 YouTube 视频：https://www.youtube.com/watch?v=fDzNJYi6Bok/。

第四部分

走进三维

"在一维中,移动一个点难道不是产生了有两个端点的线段?
在两维中,移动一段线难道不是产生了有四个端点的正方形?
在三维中,移动一个正方形难道不是产生了(我的眼睛没有看见)那种神奇的造物,立方体,有八个端点?"

——Edwin A. Abbott, Flatland: A Romance of Many Dimensions

第9章
理解 OpenGL

本项目将创建一个简单的程序，利用 OpenGL 和 GLFW 来显示纹理贴图的正方形。OpenGL 为图形处理单元（GPU）增加了一个软件接口，而 GLFW 是一个窗口工具包。我们还将学习如何使用类似 C 语言的 OpenGL 着色语言（GLSL）来编写着色器，即在 GPU 上执行的代码。着色器为 OpenGL 中的计算带来巨大的灵活性。我会展示如何用 GLSL 着色器来变换几何图形并上色，创建一个旋转的、带纹理的多边形（如图 9-1 所示）。

GPU 经过优化，以并行的方式对大量数据反复地执行相同的操作，这让它们在做这样的工作时，比中央处理单元（CPU）快得多。除了渲染计算机图形，它们也用于通用计算，现在有一些专门的语言让你用 GPU 硬件完成这样的工作。本项目将利用 GPU、OpenGL 和着色器。

Python 是一种极好的"胶水"语言。对于其他语言（如 C 语言）编写的库，有大量的 Python 绑定，让你在 Python 中使用它们。本章和第 10 章、第 11 章，将使用 PyOpenGL，即 OpenGL 的 Python 绑定，来创建计算机图形。

OpenGL 是一个"状态机"，有点像一个电器开关，有两种状态：开和关。如果从一种状态切换到另一种，开关就保持在这种新状态。然而，OpenGL 比简单的电器开关更复杂，它更像一个交换机，有许多开关和表盘。一旦更改特定设置的状态，它就保持为关，直到你打开。如果一个 OpenGL 调用绑定到某个对象上了，那随后的相关调用都会发送给这个对象，直到解除绑定。

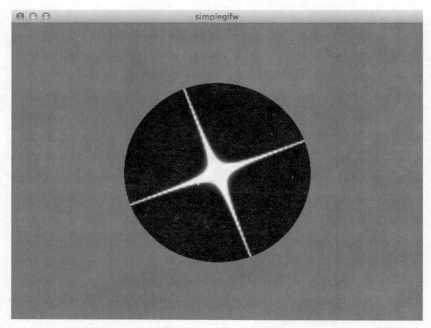

图 9-1　本章项目生成的最终图像一个旋转的多边形，它带有一个星形图案。利用着色器，这个方形的多边形边界被剪裁成一个黑色圆圈

下面是本项目引入的一些概念：
- 使用针对 OpenGL 的 GLFW 窗口库；
- 使用 GLSL 编写顶点和片段着色器；
- 进行纹理贴图；
- 使用三维变换。

首先，我们来看看 OpenGL 的工作原理。

9.1　老式 OpenGL

在大多数计算机图形系统中，绘图的方式是将一些顶点发送给处理管线，管线由一系列功能模块互相连接而成。最近，OpenGL 应用编程接口（API）从固定功能的图形管线转为可编程的图形管线。我们将专注于现代的 OpenGL，但因为在网络上会看到许多"老式"OpenGL 的例子，我会让你看看这个 API 过去的样子，以更好地感受所发生的变化。

例如，下面这个简单的老式 OpenGL 程序在屏幕上绘制一个黄色的矩形。

```
import sys
from OpenGL.GLUT import *
from OpenGL.GL import *

def display():
    glClear (GL_COLOR_BUFFER_BIT|GL_DEPTH_BUFFER_BIT)
    glColor3f (1.0, 1.0, 0.0)
    glBegin(GL_QUADS)
    glVertex3f (-0.5, -0.5, 0.0)
    glVertex3f (0.5, -0.5, 0.0)
    glVertex3f (0.5, 0.5, 0.0)
    glVertex3f (-0.5, 0.5, 0.0)
    glEnd()
    glFlush();

glutInit(sys.argv)
glutInitDisplayMode(GLUT_SINGLE|GLUT_RGB)
glutInitWindowSize(400, 400)
glutCreateWindow("oldgl")
glutDisplayFunc(display)
glutMainLoop()
```

图 9-2 展示了结果。

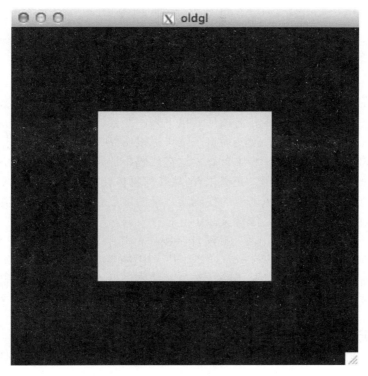

图 9-2　简单的老式 OpenGL 程序的输出

使用老式 OpenGL，要为三维图元（在这个例子中，是一个 GL_QUAD，即矩形）指定各个顶点，但随后每个顶点需要被分别发送到 GPU，这是低效的方式。这种老式编程模式伸缩性不好，如果几何图形变得复杂，程序就会很慢。对于屏幕上的顶点和像素如何变换，它只提供有限的控制（在本项目中可以看到，用新的可编程管线的范式，可以克服这些限制）。

9.2　现代 OpenGL：三维图形管线

为了让你感受现代 OpenGL 如何在较高层面上工作，让我们利用一系列的操作，即通常所谓的"三维图形管线"，在屏幕上画一个三角形。图 9-3 给出了 OpenGL 三维图形管线的简化表示。

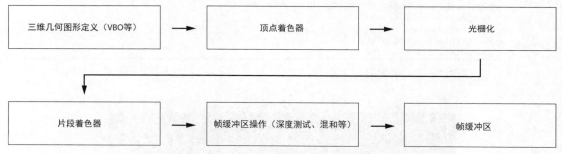

图 9-3　（简化的）OpenGL 图形管线

在第一步，通过定义在三维空间中的三角形的顶点，并指定每个顶点相关联的颜色，我们定义了三维几何图形。接下来，变换这些顶点：第一次变换将这些顶点放在三维空间中，第二次变换将三维坐标投影到二维空间。根据照明等因素，对应顶点的颜色值也在这一步中计算，这在代码中通常称为"顶点着色器"。

接着，将几何图形"光栅化"（从几何物体转换为像素），针对每个像素，执行另一个名为"片段着色器"的代码块。正如顶点着色器作用于三维顶点，片段着色器作用于光栅化后的二维像素。

最后，像素经过一系列帧缓冲区操作，其中，它经过"深度缓冲区检验"（检查一个片段是否遮挡另一个）、"混合"（用透明度混合两个片段）以及其他操作，其当前的颜色与帧缓冲区中该位置已有的颜色结合。这些变化最终体现在最后的帧缓冲区上，通常显示在屏幕上。

9.2.1　几何图元

因为 OpenGL 是一个底层图形库，你不能要求它直接画出一个立方体或球体，但建立在它之上的库可以完成这样的任务。OpenGL 只理解底层的几何图元，如点、线和三角形。

现代 OpenGL 只支持 GL_POINTS、GL_LINES、GL_LINE_STRIP、GL_LINE_LOOP、GL_TRIANGLES、GL_TRIANGLE_STRIP 和 GL_TRIANGLE_FAN 等图元类型。图 9-4 展示了图元的顶点是如何组织的。显示的每个顶点是一个三维坐标，形如(x, y, z)。

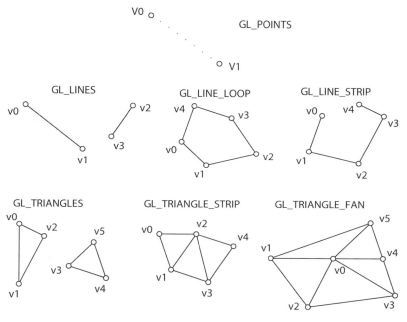

图 9-4　OpenGL 的图元

要在 OpenGL 中绘制球体，首先要在数学上定义球体的几何图形，并计算其三维顶点。然后，将这些顶点组合为基本的几何图元。例如，可以将每组三个顶点组成一个三角形。然后，可以用 OpenGL 渲染这些顶点。

9.2.2　三维变换

不学三维变换，就不懂计算机图形学。三维变换的概念都很简单易懂。你有一个物体，能对它做怎么呢？可以移动、拉伸（或挤压）或旋转。也可以做其他事情，但这 3 个任务是物体最常见的操作或变换：平移、缩放和旋转。除了这些常用的变换，还得用透视投影将三维物体映射到屏幕的二维平面。这些变换都应用于你要变换的物体上。

虽然你可能熟悉（x, y, z）形式的三维坐标，但在三维计算机图形中使用（x, y, z, w）形式的坐标，即齐次坐标（这些坐标来自数学的一个分支，即射影几何，这超出了本书的范围）。

其次坐标允许用 4×4 的矩阵来表示这些常见的三维变换。但对于这些 OpenGL 项目，你只要知道齐次坐标（x, y, z, w）相当于三维坐标(x/w, y/w, z/w, 1.0)。三维的点（1.0, 2.0, 3.0）可以用齐次坐标表示为（1.0, 2.0, 3.0, 1.0）。

下面是用变换矩阵进行三维变换的一个例子。看看矩阵乘法如何将一个点(x, y, z, 1.0)变换为(x + tx, y + ty, z + tz, 1.0)。

$$\begin{bmatrix} 1 & 0 & 0 & t_x \\ 0 & 1 & 0 & t_y \\ 0 & 0 & 1 & t_z \\ 0 & 0 & 0 & 1 \end{bmatrix} \times \begin{bmatrix} x \\ y \\ z \\ 1 \end{bmatrix} = \begin{bmatrix} x+t_x \\ y+t_y \\ z+t_z \\ 1 \end{bmatrix}$$

在 OpenGL 中经常会遇到两个术语：模型视图变换和投影变换。随着现代 OpenGL 中可定制的着色器的问世，模型视图和投影都只是普通的变换。从历史上看，在老式 OpenGL 版本中，模型视图变换用于三维模型，将它置于空间中，而投影变换将三维坐标映射到二维平面，用于显示，你会马上看到。模型视图变换是用户定义的变换，让你放置三维物体，投影变换是三维映射到二维的变换。

两种最常用的三维图形投影变换是"正投影"和"透视投影"，但这里只会用到透视投影，这是由"视角"（眼睛能看到的范围）、"近裁剪平面"（最接近眼睛的平面）、"远裁剪平面"（离眼睛最远的平面）和"纵横比"（近平面的宽度与高度之比）来确定的。这些参数共同构成了一个摄像机模型，决定了三维形状如何映射到二维屏幕上，如图 9-5 所示。图中的截头金字塔是"视景体"。"眼睛"是摄像机的三维位置（对于正投影，眼睛将在无穷远处，金字塔将变成长方体）。

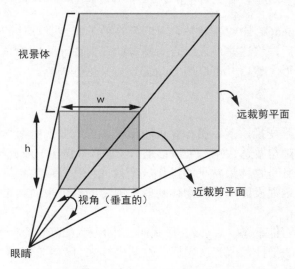

图 9-5　透视投影摄像机模型

透视投影完成后，在光栅扫描之前，图元将会被远近裁剪平面切割（或剔除），如图 9-5 所示。远近裁剪平面的选择，要让希望出现在屏幕上的三维物体位于视景体之内，否则，它们将被裁剪。

9.2.3 着色器

你已见过着色器如何融入现代 OpenGL 的可编程图形管线了。现在，我们来看看一对简单的顶点和片段着色器，了解 GLSL 的工作原理。

顶点着色器

下面是一个简单的顶点着色器：

```
❶ #version 330 core

❷ in vec3 aVert;

❸ uniform mat4 uMVMatrix;
❹ uniform mat4 uPMatrix;

❺ out vec4 vCol;

   void main() {
       // apply transformations
❻      gl_Position = uPMatrix * uMVMatrix * vec4(aVert, 1.0);
       // set color
❼      vCol = vec4(1.0, 0.0, 0.0, 1.0);
   }
```

在❶行，将着色器中使用的 GLSL 版本设置为 3.3。然后，用关键字 in 为顶点着色器定义一个输入，名为 aVert，类型为 vec3（三维向量）❷。在❸行和❹行，定义两个类型为 mat4（4×4 的矩阵）的变量，分别对应于模型视图和投影矩阵。这些变量的 uniform 前缀表明，在执行顶点着色器，对一组顶点进行指定的渲染调用时，它们不会改变。在❺行用 out 定义了顶点着色器的输出，其类型为 vec4（4D 向量，保存红、绿、蓝和 alpha 通道），是一个颜色变量。

接下来是 main() 函数，这是顶点着色器程序开始的地方。在❻行计算出 gl_Position 的值，方法是利用传入的 uniform 矩阵来变换输入的 aVert。GLSL 变量 gl_Position 用于保存变换后的顶点。在❼行，将顶点着色器的输出颜色设为红色，没有透明度，值为（1, 0, 0, 1）。管线中的下一个着色器用它作为输入。

片段着色器

现在来看一个简单的片段着色器：

```
❶ #version 330 core

❷ in vec4 vCol;
```

```
❸   out vec4 fragColor;

    void main() {
        // use vertex color
❹       fragColor = vCol;
    }
```

在❶行设置着色器的 GLSL 版本后，❷行设置 vCol 作为片段着色器的输入。这个变量 vCol 曾被设为顶点着色器输出（回忆一下，顶点着色器针对三维场景中的每个顶点执行，而片段着色器针对屏幕上的每个像素执行）。

在光栅化期间（发生在顶点着色器和片段着色器之间），OpenGL 将变换后的顶点转换为像素，通过在顶点颜色之间插值，计算顶点之间像素的颜色。

在❸行，定义了输出颜色变量 fragColor，在❹行，插值计算出的颜色被设置为输出。默认情况下，也是在大多数情况下，片段着色器的预期输出是屏幕，设置的颜色最终会出现在屏幕上（除非受到一些操作的影响，诸如发生在图形管线中最后阶段的深度测试）。

要让 GPU 执行着色器代码，需要编译和链接，成为硬件理解的指令。OpenGL 提供了一些方法来做到这一点，它报告详细的编译和链接错误，这有助于开发着色器代码。

编译过程也会为着色器中声明的变量产生一张位置（索引）表，用它可以连接 Python 代码中的变量和着色器中的变量。

9.2.4 顶点缓冲区

顶点缓冲区是 OpenGL 着色器使用的一种重要机制。现代图形硬件和 OpenGL 旨在处理大量的三维几何图形。因此，OpenGL 内建了一些机制，帮助将数据从程序传输到 GPU。程序中绘制三维几何图形的典型计划将执行以下操作：

1. 为三维几何图形的每个顶点，定义坐标、颜色和其他属性的数组；
2. 创建一个顶点数组对象（VAO），并绑定到它；
3. 为每个属性，创建顶点缓冲区对象（VBO），针对每个顶点来定义；
4. 绑定到该 VBO，并设置缓冲区数据使用预先定义的数组；
5. 指定着色器中使用的顶点属性的数据和位置；
6. 启用顶点的各种属性；
7. 渲染数据。

用顶点来定义三维几何图形之后，可以创建并绑定到一个顶点数组对象。VAO 是一种方便的方式，将几何图形分组坐标、颜色等多个数组。然后，针对每个顶点的每个属性，可以创建一个顶点缓冲区对象，并将三维数据设置给它。VBO 将顶点数据保存在 GPU 内存中。现在，剩下的就是连接缓冲区数据，以便从着色器中访问它。通过一些调用，它们使用着色器中用到的变量的位置，来实现这一点。

9.2.5 纹理贴图

接下来,看看纹理贴图:本章将使用的一种重要计算机图形技术。"纹理贴图"利用三维物体上的二维图片,让场景有逼真的感觉(就像演出的舞台背景)。纹理通常从一个图像文件中读取,被拉伸并覆盖一个几何区域,将二维坐标(范围为 [0,1])映射到多边形的三维坐标上。例如,图 9-6 展示一幅图像覆盖在立方体的一个面上(我用 GL_TRIANGLE_STRIP 图元来绘制立方体各面,顶点的顺序由面上的线来表示)。

在图 9-6 中,纹理的(0,0)角映射到立方体表面的左下顶点。类似地,可以看到纹理的其他角如何映射,最终的效果就是纹理被"粘贴"到这个立方体表面。立方体表面本身的几何图形被定义为一个三角形带,这些顶点曲折排列,从底到左上,再从底到右上。纹理是非常强大和灵活的计算机图形工具,你会在第 11 章中看到。

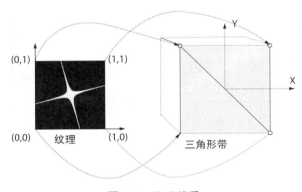

图 9-6　纹理贴图

9.2.6 显示 OpenGL

现在来谈谈如何让 OpenGL 在屏幕上绘制东西。保存所有 OpenGL 状态信息的实体叫做"OpenGL 上下文"。上下文具有一个可视的、窗口状的区域,OpenGL 的绘制在其中进行,每个进程或应用程序的每次运行可以有多个上下文,但每个线程在同一时刻只能有一个当前上下文(好在,窗口工具包将负责大多数的上下文处理)。

要让 OpenGL 的输出出现在屏幕的窗口中,就需要操作系统的帮助。对于这些项目,你将使用 GLFW,它是一个轻量级的跨平台 C 库,可以让你创建和管理 OpenGL 上下文,在窗口中显示三维图形,并处理用户的输入,如鼠标点击和按键。(附录 A 介绍了这个库的安装细节。)

因为要用 Python 写代码，而不是 C 语言，所以还要使用 GLFW 的 Python 绑定（glfw.py，在本书代码库的 common 目录中可以找到），它可以让你用 Python 访问 GLFW 的所有功能。

9.3 所需模块

我们将使用 PyOpenGL（一个流行的 OpenGL Python 绑定）来渲染，并用 numpy 数组来表示三维坐标和变换矩阵。

9.4 代码

让我们用 OpenGL 创建一个简单的 Python 应用程序。要查看完整的项目代码，请直接跳到 9.5 节。

9.4.1 创建 OpenGL 窗口

第一件事就是设置 GLFW，以便有一个 OpenGL 窗口显示渲染的结果。我创建了一个 RenderWindow 类来做这件事。

下面是这个类的初始化代码：

```
class RenderWindow:
    """GLFW Rendering window class"""
    def __init__(self):

        # save current working directory
        cwd = os.getcwd()

        # initialize glfw
❶       glfw.glfwInit()

        # restore cwd
        os.chdir(cwd)

        # version hints
❷       glfw.glfwWindowHint(glfw.GLFW_CONTEXT_VERSION_MAJOR, 3)
        glfw.glfwWindowHint(glfw.GLFW_CONTEXT_VERSION_MINOR, 3)
        glfw.glfwWindowHint(glfw.GLFW_OPENGL_FORWARD_COMPAT, GL_TRUE)
        glfw.glfwWindowHint(glfw.GLFW_OPENGL_PROFILE,
                            glfw.GLFW_OPENGL_CORE_PROFILE)

        # make a window
        self.width, self.height = 640, 480
        self.aspect = self.width/float(self.height)
❸       self.win = glfw.glfwCreateWindow(self.width, self.height,
                                         b'simpleglfw')

        # make the context current
❹       glfw.glfwMakeContextCurrent(self.win)
```

在❶行初始化 GLFW 库。然后，从❷行开始，将 OpenGL 版本设置为 OpenGL 3.3 的核心模式。在❸行，创建尺寸为 640×480 的、支持 OpenGL 的窗口。最后，在❹行，让它成为当前上下文，然后就可以进行 OpenGL 调用了。

接下来，进行一些初始化调用。

```
    # initialize GL
❶   glViewport(0, 0, self.width, self.height)
❷   glEnable(GL_DEPTH_TEST)
❸   glClearColor(0.5, 0.5, 0.5, 1.0)
```

在❶行，设置视口或屏幕尺寸（宽度和高度），OpenGL 将在其中渲染三维场景。在❷行，用 GL_DEPTH_TEST 打开深度测试。在❸行，将渲染过程中调用 glClear()时的背景颜色设置为 50%的灰色，Alpha 设置为 1.0（Alpha 是像素透明度的度量）。

9.4.2 设置回调

下一步，在 GLFW 窗口中为用户接口事件注册一些事件回调，这样就可以响应鼠标点击和按键。

```
    # set window callbacks
    glfw.glfwSetMouseButtonCallback(self.win, self.onMouseButton)
    glfw.glfwSetKeyCallback(self.win, self.onKeyboard)
    glfw.glfwSetWindowSizeCallback(self.win, self.onSize)
```

这段代码分别设置了鼠标按钮、键盘按键和窗口大小调整的回调。每当一个事件发生时，注册为回调的函数就执行。

键盘回调

来看看键盘回调：

```
    def onKeyboard(self, win, key, scancode, action, mods):
        #print 'keyboard: ', win, key, scancode, action, mods
❶       if action == glfw.GLFW_PRESS:
            # ESC to quit
            if key == glfw.GLFW_KEY_ESCAPE:
❷               self.exitNow = True
            else:
                # toggle cut
❸               self.scene.showCircle = not self.scene.showCircle
```

每次键盘事件发生时，onKeyboard()回调就被调用。该函数的参数被填充了有用的信息，如发生事件的类型（例如 key-up 或 key-down），以及按了哪个键。代码 glfw.GLFW_PRESS 是说只查找 keydown（或 PRESS）事件❶。在❷行，如果按下 ESC 键，就设置退出标志。如果按下其他任何键，翻转 showCircle 布尔值，它将传入片段着色器❸。

调整窗口大小事件

下面是调整窗口大小事件的处理：

```
def onSize(self, win, width, height):
    #print 'onsize: ', win, width, height
    self.width = width
    self.height = height
    self.aspect = width/float(height)
❶   glViewport(0, 0, self.width, self.height)
```

每次窗口大小变化时，调用 glViewport() 重置图形尺寸，确保三维场景正确绘制在屏幕上❶。同时将尺寸保存在 width 和 height 中，改变后的窗口的纵横比保存在 aspect 中。

主循环

现在来到程序的主循环（GLFW 不提供默认的程序循环）。

```
     def run(self):
         # initializer timer
❶        glfw.glfwSetTime(0)
         t = 0.0
❷        while not glfw.glfwWindowShouldClose(self.win) and not self.exitNow:
             # update every x seconds
❸            currT = glfw.glfwGetTime()
             if currT - t > 0.1:
                 # update time
                 t = currT
                 # clear
❹                glClear(GL_COLOR_BUFFER_BIT | GL_DEPTH_BUFFER_BIT)

                 # build projection matrix
❺                pMatrix = glutils.perspective(45.0, self.aspect, 0.1, 100.0)

❻                mvMatrix = glutils.lookAt([0.0, 0.0, -2.0], [0.0, 0.0, 0.0],
                                          [0.0, 1.0, 0.0])
                 # render
❼                self.scene.render(pMatrix, mvMatrix)
                 # step
❽                self.scene.step()

❾                glfw.glfwSwapBuffers(self.win)
                 # poll for and process events
❿                glfw.glfwPollEvents()
         # end
         glfw.glfwTerminate()
```

在主循环中，glfw.glfwSetTime() 将 GLFW 计时器重置为 0 ❶行。你将用这个计时器定期重新绘制图形。while 循环从❷行开始，只有在窗口关闭或 exitNow 设为 True 时退出。循环退出时，glfw.glfwTerminate() 被调用，完全关闭 GLFW。

在循环中，glfw.glfwGetTime()取得当前计时器的值❸，用它来计算自上次绘图以来的时间。这里设置期望的间隔（在这个例子中是 0.1 秒，即 100 毫秒），从而可以调整帧渲染的速度。

接着，在❹行，glClear()清除了深度和颜色缓冲区，用设置的背景颜色替换它们，准备画下一帧。在❺行，用 glutils.py（下一节将详细探讨）中定义的 perspective()方法，计算投影矩阵。这里，要求 45 度视场，近/远裁剪平面的距离为 0.1/100.0。然后，利用 glutils.py 中定义的 lookAt()方法，在❻行设置模型视图矩阵。将眼睛位置设置在（0, 0, -2)，用一个"向上"矢量(0, 1, 0)看向原点(0, 0, 0)。然后，在❼行调用 scene 对象上的 render()方法，传入这些矩阵，在❽行调用 scene.step()，以便更新所需的时间步长变量。在❾行，调用 glfwSwapBuffers()，交换前后缓冲区，从而显示更新的三维图像。❿行的 glfwPollEvents()调用检查所有 UI 事件，将控制返回给 while 循环。

> **双缓冲**
>
> 双缓冲是平滑更新屏幕上图形的渲染技术。系统维护两个缓冲区：一个前缓冲区和后缓冲区。三维渲染先被渲染到后缓冲区，完成时，前缓冲区与后缓冲区交换内容。由于缓冲区更新快速发生，这种技术产生了更流畅的视觉效果，尤其是在动画时。双缓冲和链接到 OS 的其他特征一样，是由窗口工具包（在这个例子中是 GLFW）提供的。

9.4.3 Scene 类

现在来看看 Scene 类，它负责初始化和绘制三维几何图形。

```
class Scene:
    """ OpenGL 3D scene class"""
    # initialization
    def __init__(self):
        # create shader
❶       self.program = glutils.loadShaders(strVS, strFS)

❷       glUseProgram(self.program)
```

在 Scene 类的构造函数中，先编译并加载着色器。要做到这一点，我利用了工具方法 loadShaders()❶，它定义在 glutils.py 中，为一系列的 OpenGL 调用提供了一个方便的封装，这些调用从字符串加载着色器，编译它们，并将它们链接成一个 OpenGL 程序对象。因为 OpenGL 是一个状态机，所以需要在❷行调用 glUseProgram()，设置代码使用特定的"程序对象"（因为一个项目可能有多个程序）。

现在，将 Python 代码中的变量与着色器中的变量连接起来。

```
        self.pMatrixUniform = glGetUniformLocation(self.program, b'uPMatrix')
        self.mvMatrixUniform = glGetUniformLocation(self.program, b'uMVMatrix')
```

```
# texture
self.tex2D = glGetUniformLocation(self.program, b'tex2D')
```

这段代码利用 glGetUniformLocation()方法，取得变量 uPMatrix、uMVMatrix 和 tex2D 在顶点和片段着色器中的位置。然后这些位置可以用来为着色器变量赋值。

定义三维几何图形

先为正方形定义三维几何图形。

```
            # define triangle strip vertices
❶           vertexData = numpy.array(
                [-0.5, -0.5, 0.0,
                0.5, -0.5, 0.0,
                -0.5, 0.5, 0.0,
                0.5, 0.5, 0.0], numpy.float32)

            # set up vertex array object (VAO)
❷           self.vao = glGenVertexArrays(1)
            glBindVertexArray(self.vao)
            # vertices
❸           self.vertexBuffer = glGenBuffers(1)
            glBindBuffer(GL_ARRAY_BUFFER, self.vertexBuffer)
            # set buffer data
❹           glBufferData(GL_ARRAY_BUFFER, 4*len(vertexData), vertexData,
                        GL_STATIC_DRAW)
            # enable vertex array
❺           glEnableVertexAttribArray(0)
            # set buffer data pointer
❻           glVertexAttribPointer(0, 3, GL_FLOAT, GL_FALSE, 0, None)
            # unbind VAO
❼           glBindVertexArray(0)
```

在❶行，定义三角形带的顶点数组，用于绘制正方形。设想一个以原点为中心，边长为 1.0 的正方形。该正方形的左下顶点坐标为(-0.5, -0.5, 0.0)，下一个顶点（右下）坐标为(0.5, -0.5, 0.0)，依次类推。坐标的顺序是 GL_TRIANGLE_STRIP 的顺序。在❷行，创建一个 VAO。绑定到该 VAO 后，接下来所有调用将绑定到它。在❸行，创建一个 VBO 来管理顶点数据的渲染。缓冲区绑定后，第❹行根据已定义的顶点，设置缓冲区数据。

现在，需要让着色器能访问这些数据了，这在❺行实现。GlEnableVertexAttribArray()被调用，下标为 0，因为这是你在顶点着色器中设置的顶点数据变量的位置。在❻行，glVertexAttribPointer()设置了顶点属性数组的位置和数据格式。属性的下标是 0，组件个数是 3（使用三维顶点），顶点的数据类型是 GL_FLOAT。在❼行取消 VAO 绑定，让其他的相关调用不会干扰它。在 OpenGL 中，完成工作后重置状态是最佳实践。OpenGL 是一个状态机，所以如果留下一个烂摊子，它就会一直那样。

下面的代码将图像加载为 OpenGL 纹理：

```
# texture
self.texId = glutils.loadTexture('star.png')
```

返回的纹理 ID 稍后将用于渲染。

下一步，更新 Scene 对象中的变量，让正方形在屏幕上旋转：

```
# step
def step(self):
    # increment angle
❶   self.t = (self.t + 1) % 360
    # set shader angle in radians
❷   glUniform1f(glGetUniformLocation(self.program, 'uTheta'),
                math.radians(self.t))
```

在❶行，递增角度变量 t，并利用取模操作符（%），保持该值在[0，360]的范围内。然后，在❷行利用 glUniform1f()方法，在着色器程序中设置该值。像以前一样，用 glGetUniformLocation()来获得着色器中的 uTheta 角变量的位置，而 Python 的 math.radians()方法将度转换为弧度。

现在来看看主要的渲染代码：

```
# render
def render(self, pMatrix, mvMatrix):
    # use shader
❶   glUseProgram(self.program)

    # set projection matrix
❷   glUniformMatrix4fv(self.pMatrixUniform, 1, GL_FALSE, pMatrix)

    # set modelview matrix
    glUniformMatrix4fv(self.mvMatrixUniform, 1, GL_FALSE, mvMatrix)

    # show circle?
❸   glUniform1i(glGetUniformLocation(self.program, b'showCircle'),
                self.showCircle)

    # enable texture
❹   glActiveTexture(GL_TEXTURE0)
❺   glBindTexture(GL_TEXTURE_2D, self.texId)
❻   glUniform1i(self.tex2D, 0)

    # bind VAO
❼   glBindVertexArray(self.vao)
    # draw
❽   glDrawArrays(GL_TRIANGLE_STRIP, 0, 4)
    # unbind VAO
❾   glBindVertexArray(0)
```

在❶行，设置渲染使用着色器程序。然后，从❷行开始，利用 glUniformMatrix4fv() 方法，在着色器中设置计算好的投影和模型视图矩阵。在❸行用 glUniform1i()，设置片段着色器中 showCircle 变量的当前值。OpenGL 有多纹理单元的概念，

第 9 章　理解 OpenGL　**135**

glActiveTexture()❹激活纹理单元 0（默认值）。在❺行，绑定前面生成的纹理 ID，激活它，准备渲染。在❻行，片段着色器中的 sampler2D 变量设为纹理单元 0。在❼行，绑定到先前创建的 VAO。现在你看到了使用 VAO 的好处了：实际绘制之前，不需要重复一大堆顶点缓冲相关的调用。在❽行，glDrawArrays()被调用，渲染绑定的顶点缓冲区。图元类型是一个三角形带，有四个顶点要渲染。在❾行取消绑定 VAO，这是良好的编码习惯。

定义 GLSL 着色器

现在让我们看看项目中最精彩的部分：GLSL 着色器。这是顶点着色器：

```
#version 330 core
❶ layout(location = 0) in vec3 aVert;

❷ uniform mat4 uMVMatrix;
   uniform mat4 uPMatrix;
   uniform float uTheta;

❸ out vec2 vTexCoord;

   void main() {
       // rotational transform
❹     mat4 rot = mat4(
                  vec4(cos(uTheta), sin(uTheta), 0.0, 0.0),
                  vec4(-sin(uTheta), cos(uTheta), 0.0, 0.0),
                  vec4(0.0, 0.0, 1.0, 0.0),
                  vec4(0.0, 0.0, 0.0, 1.0)
                  );
       // transform vertex
❺     gl_Position = uPMatrix * uMVMatrix * rot * vec4(aVert, 1.0);
       // set texture coordinate
❻     vTexCoord = aVert.xy + vec2(0.5, 0.5);
   }
```

在❶行，用 layout 关键字明确设置顶点的位置属性 aVert：在这个例子中设置为 0。从❷行开始，声明一些 uniform 变量：投影和模型视图矩阵和旋转角度。这些将在 Python 代码中设置。在❸行，设置一个二维矢量 vTexCoord，作为这个着色器的输出。它将作为片段着色器的输入。在着色器的 main()方法中，在❹行设立旋转矩阵，它围绕 z 轴旋转给定的角度。在❺行，利用投影、模型视图和旋转矩阵级联来计算 gl_Position。在❻行，设置一个二维向量作为纹理坐标。你可能还记得，你定义了三角形带，表示以原点为中心、边长为 1.0 的正方形。因为纹理坐标的范围是 [0,1]，所以可以通过在 x 值和 y 值上增加(0.5, 0.5)，从顶点坐标来生成它们。这也展示了着色器的计算的能力和巨大的灵活性。纹理坐标和其他变量不是神圣不可侵犯的，你可以将它们设置为任何东西。

现在，来看看片段着色器：

```
#version 330 core

❶ in vec4 vCol;
  in vec2 vTexCoord;

❷ uniform sampler2D tex2D;
❸ uniform bool showCircle;

❹ out vec4 fragColor;

  void main() {
      if (showCircle) {
          // discard fragment outside circle
❺         if (distance(vTexCoord, vec2(0.5, 0.5)) > 0.5) {
              discard;
          }
          else {
❻             fragColor = texture(tex2D, vTexCoord);
          }
      }
      else {
❼         fragColor = texture(tex2D, vTexCoord);
      }
  }
```

从❶行开始，定义了片段着色器的输入：就是设置为顶点着色器的输出变量的那些颜色和纹理坐标变量。回想一下，片段着色器基于每个像素操作，因此，对这些变量设置的值是针对当前像素的值，在整个多边形中内插。在❷行声明了 sampler2D 变量，它被连接到一个特定的纹理单元，用于查找纹理值。在❸行，声明了布尔型 uniform 标志 showCircle，这是从 Python 代码中设置的，在❹行，声明了 fragColor，作为片段着色器的输出。默认情况下，会显示在屏幕上（经过最后的帧缓冲区操作，如深度测试和混合）。

如果没有设置 showCircle 标志，在❼行，使用 GLSL 的 texture()方法来查找纹理颜色值，利用了纹理坐标和采样。实际上，你只是用星形图像来设置三角形带的纹理。但如果 showCircle 标志为真，在❺行，用 GLSL 的内置方法 distance，来检查当前像素离多边形的中心有多远。出于这个目的，它使用（内插的）纹理坐标，这是由顶点着色器传入的。如果该距离大于某一阈值（在本例中是 0.5），就调用 GLSL 的 discard 方法，丢弃当前像素。如果该距离小于阈值，就在❻行设置来自纹理的适当颜色。事实上，这样做就忽略了以正方形的中点为中心、半径为 0.5 的圆之外的像素，从而在 showCircle 设置时，切割该多边形，放入圆中。

9.5 完整代码

这个简单的 OpenGL 应用程序的完整代码分为两个文件：simpleglfw.py，包含下面展示的代码，并可以在 https://github.com/electronut/pp/tree/master/simplegl/找到；glutils.py，包括一些辅助方法，让生活更轻松，可以在 common 目录中找到。

```python
import OpenGL
from OpenGL.GL import *

import numpy, math, sys, os
import glutils

import glfw

strVS = """
#version 330 core

layout(location = 0) in vec3 aVert;

uniform mat4 uMVMatrix;
uniform mat4 uPMatrix;
uniform float uTheta;

out vec2 vTexCoord;

void main() {
    // rotational transform
    mat4 rot = mat4(
                vec4(cos(uTheta), sin(uTheta), 0.0, 0.0),
                vec4(-sin(uTheta), cos(uTheta), 0.0, 0.0),
                vec4(0.0, 0.0, 1.0, 0.0),
                vec4(0.0, 0.0, 0.0, 1.0)
                );
    // transform vertex
    gl_Position = uPMatrix * uMVMatrix * rot * vec4(aVert, 1.0);
    // set texture coordinate
    vTexCoord = aVert.xy + vec2(0.5, 0.5);
}
"""
strFS = """
#version 330 core

in vec2 vTexCoord;

uniform sampler2D tex2D;
uniform bool showCircle;

out vec4 fragColor;

void main() {
    if (showCircle) {
        // discard fragment outside circle
        if (distance(vTexCoord, vec2(0.5, 0.5)) > 0.5) {
            discard;
        }
        else {
            fragColor = texture(tex2D, vTexCoord);
        }
    }
    else {
```

```
            fragColor = texture(tex2D, vTexCoord);
        }
    }
"""

class Scene:
    """ OpenGL 3D scene class"""
    # initialization
    def __init__(self):
        # create shader
        self.program = glutils.loadShaders(strVS, strFS)

        glUseProgram(self.program)
        self.pMatrixUniform = glGetUniformLocation(self.program, b'uPMatrix')
        self.mvMatrixUniform = glGetUniformLocation(self.program, b'uMVMatrix')
        # texture
        self.tex2D = glGetUniformLocation(self.program, b'tex2D')

        # define triange strip vertices
        vertexData = numpy.array(
            [-0.5, -0.5, 0.0,
              0.5, -0.5, 0.0,
             -0.5,  0.5, 0.0,
              0.5,  0.5, 0.0], numpy.float32)

        # set up vertex array object (VAO)
        self.vao = glGenVertexArrays(1)
        glBindVertexArray(self.vao)
        # vertices
        self.vertexBuffer = glGenBuffers(1)
        glBindBuffer(GL_ARRAY_BUFFER, self.vertexBuffer)
        # set buffer data
        glBufferData(GL_ARRAY_BUFFER, 4*len(vertexData), vertexData,
                     GL_STATIC_DRAW)
        # enable vertex array
        glEnableVertexAttribArray(0)
        # set buffer data pointer
        glVertexAttribPointer(0, 3, GL_FLOAT, GL_FALSE, 0, None)
        # unbind VAO
        glBindVertexArray(0)

        # time
        self.t = 0
        # texture
        self.texId = glutils.loadTexture('star.png')

        # show circle?
        self.showCircle = False

    # step
    def step(self):
        # increment angle
        self.t = (self.t + 1) % 360
        # set shader angle in radians
```

```python
            glUniform1f(glGetUniformLocation(self.program, 'uTheta'),
                        math.radians(self.t))

    # render
    def render(self, pMatrix, mvMatrix):
        # use shader
        glUseProgram(self.program)

        # set projection matrix
        glUniformMatrix4fv(self.pMatrixUniform, 1, GL_FALSE, pMatrix)

        # set modelview matrix
        glUniformMatrix4fv(self.mvMatrixUniform, 1, GL_FALSE, mvMatrix)

        # show circle?
        glUniform1i(glGetUniformLocation(self.program, b'showCircle'),
                    self.showCircle)

        # enable texture
        glActiveTexture(GL_TEXTURE0)
        glBindTexture(GL_TEXTURE_2D, self.texId)
        glUniform1i(self.tex2D, 0)

        # bind VAO
        glBindVertexArray(self.vao)
        # draw
        glDrawArrays(GL_TRIANGLE_STRIP, 0, 4)
        # unbind VAO
        glBindVertexArray(0)

class RenderWindow:
    """GLFW Rendering window class"""
    def __init__(self):

        # save current working directory
        cwd = os.getcwd()

        # initialize glfw - this changes cwd
        glfw.glfwInit()

        # restore cwd
        os.chdir(cwd)

        # version hints
        glfw.glfwWindowHint(glfw.GLFW_CONTEXT_VERSION_MAJOR, 3)
        glfw.glfwWindowHint(glfw.GLFW_CONTEXT_VERSION_MINOR, 3)
        glfw.glfwWindowHint(glfw.GLFW_OPENGL_FORWARD_COMPAT, GL_TRUE)
        glfw.glfwWindowHint(glfw.GLFW_OPENGL_PROFILE,
                            glfw.GLFW_OPENGL_CORE_PROFILE)
        # make a window
        self.width, self.height = 640, 480
        self.aspect = self.width/float(self.height)
        self.win = glfw.glfwCreateWindow(self.width, self.height,
                                         b'simpleglfw')
```

```python
        # make context current
        glfw.glfwMakeContextCurrent(self.win)

        # initialize GL
        glViewport(0, 0, self.width, self.height)
        glEnable(GL_DEPTH_TEST)
        glClearColor(0.5, 0.5, 0.5, 1.0)

        # set window callbacks
        glfw.glfwSetMouseButtonCallback(self.win, self.onMouseButton)
        glfw.glfwSetKeyCallback(self.win, self.onKeyboard)
        glfw.glfwSetWindowSizeCallback(self.win, self.onSize)

        # create 3D
        self.scene = Scene()

        # exit flag
        self.exitNow = False


    def onMouseButton(self, win, button, action, mods):
        #print 'mouse button: ', win, button, action, mods
        pass

    def onKeyboard(self, win, key, scancode, action, mods):
        #print 'keyboard: ', win, key, scancode, action, mods
        if action == glfw.GLFW_PRESS:
            # ESC to quit
            if key == glfw.GLFW_KEY_ESCAPE:
                self.exitNow = True
            else:
                # toggle cut
                self.scene.showCircle = not self.scene.showCircle

    def onSize(self, win, width, height):
        #print 'onsize: ', win, width, height
        self.width = width
        self.height = height
        self.aspect = width/float(height)
        glViewport(0, 0, self.width, self.height)

    def run(self):
        # initializer timer
        glfw.glfwSetTime(0)
        t = 0.0
        while not glfw.glfwWindowShouldClose(self.win) and not self.exitNow:
            # update every x seconds
            currT = glfw.glfwGetTime()
            if currT - t > 0.1:
                # update time
                t = currT
                # clear
                glClear(GL_COLOR_BUFFER_BIT | GL_DEPTH_BUFFER_BIT)
```

```python
            # build projection matrix
            pMatrix = glutils.perspective(45.0, self.aspect, 0.1, 100.0)

            mvMatrix = glutils.lookAt([0.0, 0.0, -2.0], [0.0, 0.0, 0.0],
                                     [0.0, 1.0, 0.0])
            # render
            self.scene.render(pMatrix, mvMatrix)
            # step
            self.scene.step()

            glfw.glfwSwapBuffers(self.win)
            # poll for and process events
            glfw.glfwPollEvents()

        # end
        glfw.glfwTerminate()

    def step(self):
        # clear
        glClear(GL_COLOR_BUFFER_BIT | GL_DEPTH_BUFFER_BIT)

        # build projection matrix
        pMatrix = glutils.perspective(45.0, self.aspect, 0.1, 100.0)

        mvMatrix = glutils.lookAt([0.0, 0.0, -2.0], [0.0, 0.0, 0.0],
                                 [0.0, 1.0, 0.0])
        # render
        self.scene.render(pMatrix, mvMatrix)
        # step
        self.scene.step()

        glfw.SwapBuffers(self.win)
        # poll for and process events
        glfw.PollEvents()

# main() function
def main():
    print("Starting simpleglfw. "
        "Press any key to toggle cut. Press ESC to quit.")
    rw = RenderWindow()
    rw.run()
# call main
if __name__ == '__main__':
    main()
```

9.6 运行 OpenGL 应用程序

下面是该项目的运行示例：

```
$python simpleglfw.py
```

输出如图 9-1 所示。

现在，让我们快速浏览一下在 glutils.pie 中定义的一些工具方法。这个方法加载图像，作为 OpenGL 纹理：

```
def loadTexture(filename):
    """load OpenGL 2D texture from given image file"""
❶   img = Image.open(filename)
❷   imgData = numpy.array(list(img.getdata()), np.int8)
❸   texture = glGenTextures(1)
❹   glBindTexture(GL_TEXTURE_2D, texture)
❺   glPixelStorei(GL_UNPACK_ALIGNMENT, 1)
❻   glTexParameterf(GL_TEXTURE_2D, GL_TEXTURE_WRAP_S, GL_CLAMP_TO_EDGE)
    glTexParameterf(GL_TEXTURE_2D, GL_TEXTURE_WRAP_T, GL_CLAMP_TO_EDGE)
❼   glTexParameterf(GL_TEXTURE_2D, GL_TEXTURE_MAG_FILTER, GL_LINEAR)
    glTexParameterf(GL_TEXTURE_2D, GL_TEXTURE_MIN_FILTER, GL_LINEAR)
❽   glTexImage2D(GL_TEXTURE_2D, 0, GL_RGBA, img.size[0], img.size[1],
                 0, GL_RGBA, GL_UNSIGNED_BYTE, imgData)
    return texture
```

在❶行，loadTexture()函数用 Python 图像库（PIL）的 Image 模块读取图像文件。然后在❷行，获取 Image 对象的数据，放入 8 位的 numpy 数组，在❸行，创建一个 OpenGL 纹理对象，这是在 OpenGL 中利用纹理做任何事的先决条件。在❹行，执行现在你比较熟悉的绑定 texture 对象，这样所有后来纹理相关的设置都应用于该对象。在❺行，将数据的拆包对齐设置为 1，这意味着该图像数据被硬件认为是 1 字节（或 8 位）的数据。从❻行开始，告诉 OpenGL 如何处理边缘的纹理。在这个例子中，在几何图形的边缘截取纹理颜色（指定纹理坐标时，惯例是使用字母 S 和 T 表示轴，而不是 x 和 y）。在❼行和下一行，指定插值类型，在拉伸或压缩纹理来覆盖多边形时采用。在这个例子中，指定为"线性滤波"。在❽行，设置绑定纹理中的图像数据。此时，图像数据传送到显存，纹理准备好使用了。

9.7 小结

恭喜你完成了使用 Python 和 OpenGL 的第一个程序。你已踏上进入三维图形编程迷人世界的旅程。

9.8 实验

下面有一些修改这个项目的想法。

1. 这个项目中，顶点着色器围绕 z 轴（0，0，1）旋转正方形。你能让它绕轴（1，1，0）旋转吗？可以用两种方式实现：第一，修改着色器中的旋转矩阵，第二，在 Python 代码计算这个矩阵，将它作为一个 uniform 传入着色器。两种方法都试一下。

2．这个项目中，纹理坐标在顶点着色器内产生，并传入片段着色器。这是一种方法，它有效只是因为三角形带的顶点选择了方便的值。请将纹理坐标作为单独的属性传入顶点着色器，类似于顶点传入的方式。现在，你能让星形纹理平铺三角形带吗？不是显示一颗星，而是在正方形上生成 4×4 的星星网格（提示：使用大于 1.0 的纹理坐标，并将 glTexParameterf()中的 GL_TEXTURE_WRAP_S/T 参数设置为 GL_REPEAT）。

3．只修改片段着色器，能让你的正方形如图 9-7 所示（提示：使用 GLSL 的 sin()函数）？

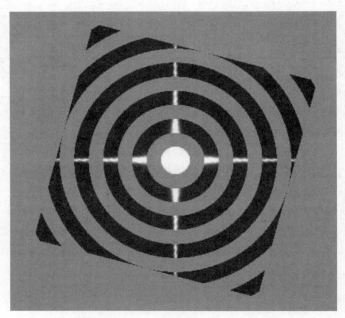

图 9-7　使用片段着色器来画出同心圆

第 10 章
粒子系统

在计算机图形世界中,粒子系统是用许多小图元(如点、线、三角形和多边形)来表示的物体,如烟雾、火焰,甚至头发,没有明确的几何形状,因此很难用标准技术来建模。

例如,如何在计算机上制造一次爆炸效果?设想爆炸从空间中的一个点开始,然后向外扩张,作为一个快速扩大的、复杂的三维实体,随时间而改变形状和颜色。不夸张地说,尝试建立数学模型就令人望而生畏。

但现在设想一下,爆炸包含一群细小的粒子,每个粒子有关联的位置和颜色。爆炸开始时,粒子在空间中的一个点上聚成一图。随着时间的推移,它们根据一定的数学规则向外移动,并改变颜色,让你定期绘制所有粒子,从而生成爆炸的动画。利用好的数学模型、大量粒子,以及透明度和公告板(billboarding)这样的渲染技术,可以创建逼真的效果,如图 10-1 所示。

本项目会制定粒子运动的数学模型,将它表示为时间的函数,并利用图形处理单元(GPU)的着色器进行计算。然后,会设计一种渲染方案,利用一种名为公告板的技术,它让二维图像一直面向观众,从而使二维图像看起来像是三维的,用一种令人信服的方式来绘制这些粒子。还会用 OpenGL 着色器让粒子旋转,并生成动画场景。你可以通过按键来打开或关闭各种效果,进行比较。

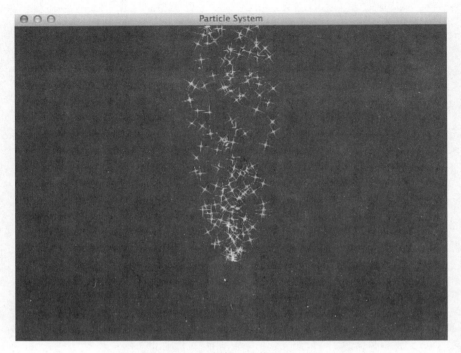

图 10-1　已完成项目的运行示例

数学模型将设置每个粒子的初始位置和速度,并决定粒子如何随时间运动。你可以让每个纹理的黑色区域透明,利用正方形图像创建火花,保持每个火花面向观众,使它们看起来有立体感。你会生成粒子的动画,定期更新它们的位置,其亮度随着时间的推移逐渐减弱。

下面是将要探索的一些概念:
- 制定喷泉粒子系统的数学模型;
- 利用 GPU 着色器计算;
- 利用纹理和公告板模拟复杂的三维对象;
- 利用 OpenGL 渲染功能,如混合、深度遮掩和 Alpha 通道,绘制半透明物体;
- 利用相机模型绘制三维透视图。

10.1　工作原理

要创建动画,就需要一个数学模型。从一个固定点开始,移动一些粒子,随时间推移沿抛物线轨迹运动,形成火花喷泉,如图 10-2 所示。这就是喷泉粒子系统。

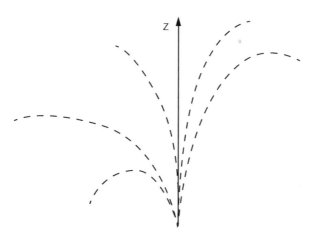

图 10-2 5 个示例火花在喷泉粒子系统中的轨迹

我们的粒子系统应具有以下特性：
- 粒子应该从固定点的出现，它们的运动轨迹应该是抛物线；
- 粒子应该能相对于喷泉的垂直轴（z 轴或高度），运动预定距离；
- 较接近喷泉垂直轴中心的粒子，应该比离中心较远的粒子初速度更大；
- 要产生更逼真的效果，粒子不应全部同时喷射；
- 粒子的亮度应该随时间推移逐渐淡出。

10.1.1 为粒子运动建模

假设粒子系统包含 N 个粒子。第 i 个粒子的运动方程如下：

$$P_t^i = P_0 + V_0^i t + \frac{1}{2}at^2$$

这里，P_t 是在时间 t 的位置。P_0 为粒子的初始位置，V_0 是粒子的初始速度，a 是加速度。可以认为 a 是系统中的重力加速度，让粒子沿向下的弧线运动。

这些参数都是三维向量，可以表示为三维坐标。例如，使用加速度值是（0，0，–9.8），这是地球在 z 轴（垂直）方向的重力加速度，单位是米每平方秒。

10.1.2 设置最大范围

要让喷泉看起来逼真，粒子应相对于圆锥体的 z 轴，以不同的角度飞出。但也要设置一个最大范围，让每个粒子的初始速度在一定范围内，让这些粒子呈漏斗状，形成喷泉的样子。为了实现这个目标，将速度与垂直轴的最大角度确定为 20 度，如图 10-3 所示。

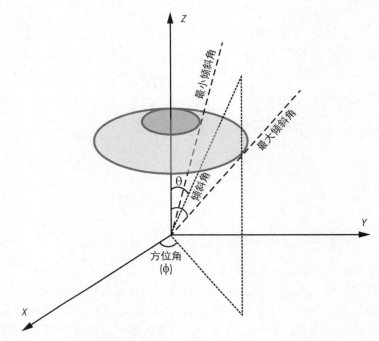

图 10-3 限制每个粒子的初始速度范围。每个粒子分配的速度在阴影圆内

在图 10-3 中，方位角 ϕ 是速度向量与 x 轴的夹角。倾斜角 θ 是速度向量与 z 轴的夹角。倾斜角的选择范围让速度向量处于图中浅灰色阴影区域内。

速度方向应该从大圆圈包围的半球部分中随机选择。这些方向的端点处于一个单位半径的球面上，这样就可以用球坐标系来计算它们。此外，我们希望速度值随着与轴的夹角增大而减小。考虑到这一点，粒子的初始速度如下：

$$V_0^i = (1-\alpha^2)V$$

这里，α 是粒子的倾斜角与最大角度（这个例子中是 20 度）之比。因此，随着这个比值逐渐变为 1.0，速度（二次地）下降。速度 V 是单位球面上一个点，像下面这样：

$$V = (\cos(\theta)\sin(\Phi), \sin e(\theta)\sin(\Phi), \cos(\Phi))$$

在[0，20]度范围内随机选择一个倾斜角，在[0，360]度范围内随机选择一个方位角：

$$\theta = \text{random}([0, 20]), \Phi = \text{random}([0, 360])$$

这个方程让粒子指向图 10-3 中圆圈区域内的一个随机点（注意，在这个程序中，所有角度计算需要用弧度，不是度）。

我们也希望确保粒子不在同一时间开始（原因请参阅 10.10 节）。要做到这一点，为每个粒子计算一个时间延迟，稍后在计算粒子位置时使用。第 i 个粒子的延迟计算如下：

$$t_{\text{tag}}^i = 0.05i$$

这些方程中随意采用的数值常数是怎么来的？例如，为什么用 0.05 作为延迟时间，用 20 度作为最大角度？答案是实验。重要的是创建一个基本模型，然后调整这些常数，以获得最佳视觉效果。改变程序中的这些参数，看看不同的值如何影响结果。

10.1.3 渲染粒子

渲染粒子的一种简单方式，就是将它们绘制为点。OpenGL 有 GL_POINTS 图元，本质上是屏幕上的一个点；你可以控制点的像素大小和颜色。但我们希望粒子看起来像小火花，并且随着喷发而旋转。

从头开始绘制火花太复杂了，所以需要一张火花的照片，作为纹理粘贴到一个矩形（也称为"四边形"）。喷泉中的每个粒子都被绘制成火花的三角形纹理图像。但是，这提出两个问题。首先，不希望正方形火花，因为那显得很假。其次，如果从其他角度看喷泉，四边形的方向会不对。

10.1.4 利用 OpenGL 混合来创建更逼真火花

为了创造更逼真的火花，我们利用 OpenGL 的混合（blending）。结合传入的片段（片段着色器执行之后）和帧缓冲区中已有的内容。这项操作通常涉及 Alpha 通道。

例如，假设在屏幕上绘制两个多边形，并想将它们混合在一起。你可以用 alpha 混合的技巧，它的工作原理是将两个透明片重叠起来。alpha 通道表示像素的不透明度，它是透明程度的量度。除了表示一个像素颜色的红、绿、蓝分量之外，还可以存储一个 alpha 值，得到的颜色方案称为 RGBA。对于 32 位的 RGBA 颜色方案，alpha 值的范围是 [0,255]，0 是完全透明，255 是完全不透明。alpha 通道本身不做任何事情。只有用 alpha 值来改变像素的最终 RGB 值时，才会创建各种透明效果。

OpenGL 提供了几种方法来定制混合方程。对于喷泉粒子，我们将使用图 10-4 所示的纹理，但要让纹理的黑色区域消失，这样就会只看到火花。

启用 OpenGL 的混合，用片段的 alpha 值乘以纹理的颜色，可以让黑色区域消失。对于黑色区域，RGB 颜色值是（0，0，0），如果乘以 alpha 值，将得到 0。因此，经过混合，黑色区域在最终图像中的不透明度是 0，你看到的只是背景颜色，实际上去掉了火花纹理的黑色区域（alpha 值在片段着色器中设置，在 10.3.5 节中有介绍）。

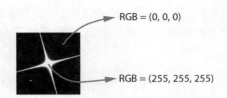

图 10-4　火花纹理，标有 RGB 值（0，0，0）为黑色，（255，255，255）为白色

> 注意　除了在着色器中设置 alpha 值，也可以利用纹理图像中的 alpha 通道，为黑色和白色区域设定不同的 alpha 值，从而控制透明度。在着色器中使用 alpha 值，创建带有黑色背景的纹理，再让所有黑色区域透明，这更简单，优于在纹理中使用 alpha 通道，让特定区域有不同的透明度值。

10.1.5　使用公告板

对于第二个问题（如果从其他角度看喷泉，四边形的方向不对），我们将使用公告板。我们不是绘制复杂的三维物体，而是放置一个二维的图片，让你总是看到其正面（面向观看的方向），作为一种公告板。例如，在开发三维游戏时，背景是树木景观，可以用一个带纹理的多边形公告板来替代树木景观。只要玩家靠得不太近，假的"图片树"看起来就很逼真。

现在来看看放置多边形背后的数学，以便让它总是面向观看的方向。图 10-5 展示了如何将一个纹理四边形变成一个公告板。

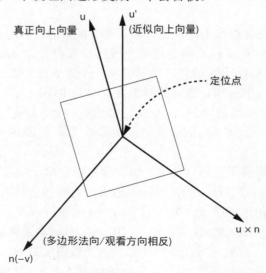

图 10-5　公告板

对准是相对于四边形上的定位点来说的。在这个例子中，会选择四边形的中心。要对准四边形，需要三个正交向量，创建一个小坐标系。我们感兴趣的第一个向量是 n，代表四边形的法线向量。法线向量垂直于四边形所在的平面，这意味着设定的法线向量方向，就是四边形面向的方向。我们希望四边形面向观看的方向，即 v，所以 n 向量需要对准该向量，但方向相反。因为观看的方向朝向屏幕，所以四边形的法线向量应该指向相反的方向：从屏幕向外。

这意味着我们希望方向 $n = -v$。现在，选择一个向量 u，这是四边形最终位置的近似向上向量。选择 u 为（0，0，1），因为 z 方向指"向上"（说这个向量是近似的，是因为虽然我们事先知道四边形的法线向量，但不知道摄像头的方向）。然后计算第三个向量 r，这里 $r = u \times n$（两个向量的叉积）。现在有了两个正交向量，n 和 r，都在四边形的平面中。叉积这两个向量，得到一个新向量。因此，得到了新的向上向量 $u' = n \times r = n \times (u \times n)$。我们需要一个新的向上向量，以确保这 3 个向量是正交的，或彼此垂直。最后，在计算过程中，这些向量都需要归一化，让长度等于一个单位，从而创建一个正交坐标系（其中所有向量是单位长度并且正交）。有了这 3 个向量，就可以根据三维图形理论，用一个旋转矩阵进行任意方向的旋转。旋转矩阵 R 将位于原点的正交坐标系旋转到 r、u' 和 n 构成的坐标系：

$$R = \begin{bmatrix} r_x & u'_x & n_x & 0 \\ r_y & u'_y & n_y & 0 \\ r_z & u'_z & n_z & 0 \\ 0 & 0 & 0 & 1 \end{bmatrix}$$

应用这个旋转矩阵，将带纹理的四边形正确地对准观看的方向，使其成为一个公告板。

10.1.6 生成火花动画

要生成火花喷泉的动画，就要利用 GLFW 库，通过更新渲染和时间，定时绘制粒子系统的各个位置。

10.2 所需模块

我们将用 PyOpenGL（一个流行的 Python OpenGL 绑定）来渲染，用 numpy 的数组来表示三维坐标变换矩阵。

10.3 粒子系统的代码

我们开始先定义喷泉中使用的粒子的三维几何形状。然后，再看看如何创建动画中粒子之间的时间延迟，如何为粒子设置初始速度，以及如何在程序中使用 OpenGL 的顶点和片段着色器。最后，看看如何将所有这些整合起来，渲染粒子系统。整个项目的代码，请直接跳到 10.4 节。

喷泉的代码封装在一个类中，名为 ParticleSystem，它创建了粒子系统，建立了 OpenGL 着色器，用 OpenGL 渲染系统，每隔五秒钟重新开始动画。

10.3.1 定义粒子的几何形状

首先，创建一个顶点数组对象（VBO）来管理后续的顶点属性数组，从而定义这些粒子的几何形状。

```
# create Vertex Array Object (VAO)
self.vao = glGenVertexArrays(1)
# bind VAO
glBindVertexArray(self.vao)
```

每个粒子是一个正方形，其顶点和纹理坐标定义如下：

```
        # vertices
        s = 0.2
❶       quadV = [
        -s, s, 0.0,
        -s, -s, 0.0,
        s, s, 0.0,
        s, -s, 0.0,
        s, s, 0.0,
        -s, -s, 0.0
        ]
❷ vertexData = numpy.array(numP*quadV, numpy.float32)
❸ self.vertexBuffer = glGenBuffers(1)
❹ glBindBuffer(GL_ARRAY_BUFFER, self.vertexBuffer)
❺ glBufferData(GL_ARRAY_BUFFER, 4*len(vertexData), vertexData,
               GL_STATIC_DRAW)

  # texture coordinates
❻ quadT = [
    0.0, 1.0,
    0.0, 0.0,
    1.0, 1.0,
    1.0, 0.0,
    1.0, 1.0,
    0.0, 0.0
    ]
  tcData = numpy.array(numP*quadT, numpy.float32)
  self.tcBuffer = glGenBuffers(1)
  glBindBuffer(GL_ARRAY_BUFFER, self.tcBuffer)
  glBufferData(GL_ARRAY_BUFFER, 4*len(tcData), tcData, GL_STATIC_DRAW)
```

在❶行，定义了正方形的顶点，各边以原点为中心，长度为 0.4。对于两个 GL_TRIANGLES，顶点的排序是相同的。然后，在❷行，将这些顶点重复 numP 次，创建一个 numpy 数组：每个四边形表示系统中的一个粒子（所有要绘制的几何图形都放入到一个大数组）。

接下来，将这些顶点放入一个顶点缓冲区对象，就像在第 9 章中所做的一样。在❸行创建 VBO，在❹行绑定。然后在❺行，用顶点数据填充绑定的缓冲区。代码 4 * len(vertexData) 指明在 vertexData 数组中，每个元素需要 4 个字节。

最后，在❻行定义四边形的纹理坐标，随后几行代码设置了相关的 VBO。

10.3.2　为粒子定义时间延迟数组

接下来，为粒子定义时间延迟数组。我们希望每组四个顶点的时间延迟一样，这代表一个正方形的粒子，如以下代码所示：

```
                # time lags
❶               timeData = numpy.repeat(0.005*numpy.arange(numP, dtype=numpy.float32),
                            4)
                self.timeBuffer = glGenBuffers(1)
                glBindBuffer(GL_ARRAY_BUFFER, self.timeBuffer)
                glBufferData(GL_ARRAY_BUFFER, 4*len(timeData), timeData,
                            GL_STATIC_DRAW)
```

在❶行，numpy.arange() 创建了一个数组，包含不断增加的值，即 [0，1，...，numP-1]。将这个数组乘以 0.005，用参数 4 调用 numpy.repeat()，产生一个数组，即 [0.0，0.0，0.0，0.0，0.005，0.005，0.005，0.005，...]。接下来的代码设置了 VBO。

10.3.3　设置粒子初始速度

接下来生成粒子的初始速度。我们的目标是生成一些随机的速度，它们与垂直轴的夹角不超过某个最大值。下面是代码：

```
                # velocites
                velocities = []
                # cone angle
❶               coneAngle = math.radians(20.0)
                # set up particle velocities
                for i in range(numP):
                    # inclination
❷                   angleRatio = random.random()
                    a = angleRatio*coneAngle
                    # azimuth
❸                   t = random.random()*(2.0*math.pi)
                    # get velocity on sphere
❹                   vx = math.sin(a)*math.cos(t)
                    vy = math.sin(a)*math.sin(t)
                    vz = math.cos(a)
                    # speed decreases with angle
❺                   speed = 15.0*(1.0 - angleRatio*angleRatio)
                    # add a set of calculated velocities
❻                   velocities += 6*[speed*vx, speed*vy, speed*vz]
                # set up velocity vertex buffer
                self.velBuffer = glGenBuffers(1)
                glBindBuffer(GL_ARRAY_BUFFER, self.velBuffer)
❼               velData = numpy.array(velocities, numpy.float32)
                glBufferData(GL_ARRAY_BUFFER, 4*len(velData), velData, GL_STATIC_DRAW)
```

在❶行，定义了限制喷泉粒子轨迹的锥角。（注意，用内置 math.radians()方法，将角度转换为弧度。）接下来，用 10.1.1 节中讨论的公式，计算每个粒子的速度。

在❷行，生成一个随机分数，后面一行用它乘以最大倾斜角，来计算当前倾斜角。接着，在❸行生成方位角，因为 random.random()返回值在[0，1]之间，所以用该值乘以 2.0 * math.pi，得到 0 至 2 的一个随机弧度。

从❹行开始，利用球坐标公式，计算单位球面上的速度向量。在❺行，用❷行得到的角度比算出一个速度，它与垂直角度成反比。在❻行，计算粒子的最终速度，并针对两个三角形的所有 6 个顶点，重复这个值。在❼行，利用 Python 列表创建一个 numpy 的数组，这就可以为这些速度创建一个 VBO 了。最后，启用所有的顶点属性，并为顶点缓冲区设置数据格式（因为这个过程类似于第 9 章，这里略过，直接讨论顶点着色器）。

10.3.4　创建顶点着色器

顶点着色器处理各个顶点，从而计算粒子系统的轨迹。下面是它的代码：

```
#version 330 core

in vec3 aVel;
in vec3 aVert;
in float aTime0;
in vec2 aTexCoord;

uniform mat4 uMVMatrix;
uniform mat4 uPMatrix;
uniform mat4 bMatrix;
uniform float uTime;
uniform float uLifeTime;
uniform vec4 uColor;
uniform vec3 uPos;

out vec4 vCol;
out vec2 vTexCoord;
```

顶点着色器对四边形的每个顶点执行，对粒子系统中的所有四边形。它首先定义属性数组中的一些变量，它代表要传入 VBO 的数组。然后定义一些 uniform 变量，它们在执行着色器时保持不变。最后定义 out 数量，它们在顶点着色器中设置，并传递给片段着色器进行插值。

现在，来看看着色器的 main()函数：

```
void main() {
    // set position
❶   float dt = uTime - aTime0;
❷   float alpha = clamp(1.0 - 2.0*dt/uLifeTime, 0.0, 1.0);
❸   if(dt < 0.0 || dt > uLifeTime || alpha < 0.01) {
        // out of sight!
        gl_Position = vec4(0.0, 0.0, -1000.0, 1.0);
```

```glsl
        }
        else {
            // calculate new position
❹          vec3 accel = vec3(0.0, 0.0, -9.8);
            // apply a twist
            float PI = 3.14159265358979323846264;
❺          float theta = mod(100.0*length(aVel)*dt, 360.0)*PI/180.0;
❻          mat4 rot = mat4(vec4(cos(theta), sin(theta), 0.0, 0.0),
                vec4(-sin(theta), cos(theta), 0.0, 0.0),
                vec4(0.0, 0.0, 1.0, 0.0),
                vec4(0.0, 0.0, 0.0, 1.0));
            // apply billboard matrix
❼          vec4 pos2 = bMatrix*rot*vec4(aVert, 1.0);
            // calculate position
❽          vec3 newPos = pos2.xyz + uPos + aVel*dt + 0.5*accel*dt*dt;
            // apply transformations
❾          gl_Position = uPMatrix * uMVMatrix * vec4(newPos, 1.0);
        }
        // set color
❿      vCol = vec4(uColor.rgb, alpha);
        // set texture coordinates
        vTexCoord = aTexCoord;
}
```

在❶行，针对特定的粒子计算当前经过的时间，这是当前时间步骤与该粒子的延迟时间之间的差。然后在❷行为顶点计算 alpha 值，该值随着时间的流逝而减小，让粒子逐渐淡出。利用 GLSL 中的 clamp()，将值限制在范围[0，1]之内。

为了在粒子生命周期结束时让粒子消失，将它们放在 OpenGL 的视锥之外，这样它们会被裁剪掉。在❸行，检查粒子的生命周期是否结束（根据粒子系统构造函数中的设置），或者其 alpha 值低于特定值，这时将最终位置设置到视锥之外。

在❹行，设置粒子加速度为 9.8 米/秒 ^{2}，即由于地球引力而产生的加速度。当粒子以较高的初始速度飞出喷泉时，为了让它们快速旋转，使用 mod()方法（类似于 Python 的取模运算%），在❺行，将角度值限制在范围[0，360]之内。在❻行，利用这个计算出的角度，绕四边形的 z 轴旋转，根据绕 z 轴旋转角度的变换矩阵公式，像下面这样：

$$R_{\theta,z} = \begin{bmatrix} \cos(\theta) & \sin(\theta) & 0.0 & 0.0 \\ -\sin(\theta) & \cos(\theta) & 0.0 & 0.0 \\ 0.0 & 0.0 & 0.0 & 0.0 \\ 0.0 & 0.0 & 0.0 & 1.0 \end{bmatrix}$$

在❼行，对粒子的顶点应用两个转换：刚刚计算出的旋转和用 bMatrix 的公告板旋转。在❽行，利用前面讨论的运动方程，计算顶点的当前位置（该行中的 uPos 只是让你任意指定喷泉原点的位置）。

在❾行，对粒子位置应用模型视图和投影矩阵。最后，在❿行和下一行中，设置的颜色和纹理坐标，传入片段着色器用于插值（请注意，在❷行基于计算设置顶点的 alpha 值）。

10.3.5　创建片段着色器

现在，来看看片段着色器，它设置像素的颜色。

```
#version 330 core

uniform sampler2D uSampler;
in vec4 vCol;
in vec2 vTexCoord;
out vec4 fragColor;

void main() {
    // get the texture color
❶   vec4 texCol = texture2D(uSampler, vec2(vTexCoord.s, vTexCoord.t));
    // multiply texture color by set vertex color; use the vertex color alpha
❷   fragColor = vec4(texCol.rgb*vCol.rgb, vCol.a);
}
```

在❶行，用 GLSL 的 texture2D()方法，利用从顶点着色器传入的纹理坐标，为火花图像查找基础纹理颜色（从用作纹理的图像中查找的颜色）。然后，将这个纹理颜色值乘以火花喷泉的颜色，并将得到的颜色设置给输出变量 fragColor❷。喷泉的颜色是在每个粒子系统重新启动时，由 ParticleSystem 的 restart()方法随机设置的。alpha 值是根据顶点着色器中的计算来设置的，并在渲染时用于混合。

10.3.6　渲染

代码按照以下步骤来渲染喷泉粒子系统：

1. 启用顶点/片断程序；
2. 设置模型视图和投影矩阵；
3. 基于当前摄像机的观看方向，计算并设置公告板矩阵；
4. 针对位置、时间、生命周期和颜色，设置 uniform 变量；
5. 针对顶点、纹理坐标、时间延迟和速度，设置顶点属性数组；
6. 启用纹理并绑定到粒子火花纹理；
7. 禁用深度缓冲写入；
8. 启用 OpenGL 的混合；
9. 绘制几何图形。

让我们来看看实现其中一些步骤的代码片段。

计算公告板的旋转矩阵

这段代码计算公告板的旋转矩阵：

```
           N = camera.eye - camera.center
❶          N /= numpy.linalg.norm(N)
           U = camera.up
           U /= numpy.linalg.norm(U)
           R = numpy.cross(U, N)
           U2 = numpy.cross(N, R)
❷          bMatrix = numpy.array([R[0], U2[0], N[0], 0.0,
                                  R[1], U2[1], N[1], 0.0,
                                  R[2], U2[2], N[2], 0.0,
                                  0.0, 0.0, 0.0, 1.0], numpy.float32)
❸          glUniformMatrix4fv(self.bMatrixU, 1, GL_TRUE, bMatrix)
```

你已经看到了构造一个旋转矩阵背后的理论，该矩阵让四边形保持为"公告板"，即对准观看方向（这是必需的，以便让喷泉粒子总是面对观察的方向）。在❶行，用 numpy.linalg.norm()对向量规一化（这使得向量的大小等于 1）。在❷行，将旋转矩阵组装为一个 numpy 数组，然后在❸行将它放入程序中。

主要的渲染代码

主要的渲染代码用 alpha 混合，让粒子系统有透明性。这种技术在 OpenGL 中常用于渲染半透明物体。

```
           # enable texture
❶          glActiveTexture(GL_TEXTURE0)
❷          glBindTexture(GL_TEXTURE_2D, self.texid)
❸          glUniform1i(self.samplerU, 0)

           # turn depth mask off
           if self.disableDepthMask:
❹              glDepthMask(GL_FALSE)

           # enable blending
           if self.enableBlend:
❺              glBlendFunc(GL_SRC_ALPHA, GL_ONE)
❻              glEnable(GL_BLEND)

           # bind VAO
❼          glBindVertexArray(self.vao)
           # draw
❽          glDrawArrays(GL_TRIANGLES, 0, 6*self.numP)
```

在❶行，激活第一个 OpenGL 纹理单元（GL_TEXTURE0）。我们只有一个纹理单元，但由于在 OpenGL 上下文中，同一时间可以激活多个纹理单元，所以针对每个纹理单元显式调用是良好的编程习惯。在❷行，激活纹理对象，它是在 ParticalSystem 的构造函数中，利用 glutils.loadTexture()和火花图像创建的。

纹理利用采样在着色器中访问，在❸行，设置采样变量使用第一个纹理单元 GL_TEXTURE0。然后，利用 OpenGL 混合切除纹理中的黑色像素，但这些"看不见"的像素仍然具有关联的深度值，可以掩盖那些在它们后面的、其他粒子的某些部分。为了避免这种情况，在❹行禁用深度缓存写入。

注意　　严格来说，这是错误的绘制方式，因为如果混合这些半透明的物体和不透明的物体，它们的深度测试就不正确。渲染这样的场景，正确的做法应该是先绘制不透明的物体，然后启用混合，按深度从后到前对半透明物体排序，最后绘制它们。但是，因为有这么多运动的粒子，这个简单的近似是可以接受的，而且最后，它看起来不错，而这才是你关心的。

在❺行，设置 OpenGL 混合功能，以便使用从片段着色器传入的源像素的 alpha 值，在❻行，启用 OpenGL 混合。然后，在❼行绑定到创建的 VAO，这启用了你建立的所有的顶点属性，在❽行，在屏幕上绘制绑定的顶点缓冲区对象。

10.3.7　Camera 类

最后，Camera 类设置了 OpenGL 的观看参数：

```
# a simple camera class
class Camera:
    """helper class for viewing"""
❶   def __init__(self, eye, center, up):
        self.r = 10.0
        self.theta = 0
        self.eye = numpy.array(eye, numpy.float32)
        self.center = numpy.array(center, numpy.float32)
        self.up = numpy.array(up, numpy.float32)

    def rotate(self):
        """rotate eye by one step"""
❷       self.theta = (self.theta + 1) % 360
        # recalculate eye
❸       self.eye = self.center + numpy.array([
                self.r*math.cos(math.radians(self.theta)),
                self.r*math.sin(math.radians(self.theta)),
                0.0], numpy.float32)
```

三维立体图的特征通常在于 3 个参数：眼睛位置、向上向量和方向向量。Camera 类包含这些参数，并提供了一种便捷的方式，在每一个时间步骤旋转。

在❶行的构造函数设置了 camera 对象的初始值。调用 rotate()方法时，增加旋转角度❷，并计算旋转后新的眼睛位置和方向❸。

注意　　点 $(r\cos(\theta), r\sin(\theta))$ 表示一个点，它在以原点为中心、半径为 r 的圆上，θ 是该点到原点的连线与 x 轴的夹角。转换中使用了 center，确保即使旋转中心不在原点，也能工作。

10.4　粒子系统完整代码

这是粒子系统的完整代码。也可以在 https://github.com/electronut/pp/tree/master/

particle-system/ps.py 找到它。

```python
import sys, random, math
import OpenGL
from OpenGL.GL import *
import numpy
import glutils

strVS = """
#version 330 core

in vec3 aVel;
in vec3 aVert;
in float aTime0;
in vec2 aTexCoord;

uniform mat4 uMVMatrix;
uniform mat4 uPMatrix;
uniform mat4 bMatrix;
uniform float uTime;
uniform float uLifeTime;
uniform vec4 uColor;
uniform vec3 uPos;

out vec4 vCol;
out vec2 vTexCoord;

void main() {
    // set position
    float dt = uTime - aTime0;
    float alpha = clamp(1.0 - 2.0*dt/uLifeTime, 0.0, 1.0);
    if(dt < 0.0 || dt > uLifeTime || alpha < 0.01) {
        // out of sight!
        gl_Position = vec4(0.0, 0.0, -1000.0, 1.0);
    }
    else {
        // calculate new position
        vec3 accel = vec3(0.0, 0.0, -9.8);
        // apply a twist
        float PI = 3.14159265358979323846264;
        float theta = mod(100.0*length(aVel)*dt, 360.0)*PI/180.0;
        mat4 rot = mat4(vec4(cos(theta), sin(theta), 0.0, 0.0),
                vec4(-sin(theta), cos(theta), 0.0, 0.0),
                vec4(0.0, 0.0, 1.0, 0.0),
                vec4(0.0, 0.0, 0.0, 1.0));
        // apply billboard matrix
        vec4 pos2 = bMatrix*rot*vec4(aVert, 1.0);
        // calculate position
        vec3 newPos = pos2.xyz + uPos + aVel*dt + 0.5*accel*dt*dt;
        // apply transformations
        gl_Position = uPMatrix * uMVMatrix * vec4(newPos, 1.0);
    }
    // set color
    vCol = vec4(uColor.rgb, alpha);
    // set tex coords
    vTexCoord = aTexCoord;
}
"""

strFS = """
```

```glsl
#version 330 core

uniform sampler2D uSampler;
in vec4 vCol;
in vec2 vTexCoord;
out vec4 fragColor;

void main() {
    // get texture color
    vec4 texCol = texture(uSampler, vec2(vTexCoord.s, vTexCoord.t));
    // multiply by set vertex color; use the vertex color alpha
    fragColor = vec4(texCol.rgb*vCol.rgb, vCol.a);
}
"""

# a simple camera class
class Camera:
    """helper class for viewing"""
    def __init__(self, eye, center, up):
        self.r = 10.0
        self.theta = 0
        self.eye = numpy.array(eye, numpy.float32)
        self.center = numpy.array(center, numpy.float32)
        self.up = numpy.array(up, numpy.float32)

    def rotate(self):
        """rotate eye by one step"""
        self.theta = (self.theta + 1) % 360
        # recalculate eye
        self.eye = self.center + numpy.array([
                self.r*math.cos(math.radians(self.theta)),
                self.r*math.sin(math.radians(self.theta)),
                0.0], numpy.float32)

# particle system class
class ParticleSystem:

    # initialization
    def __init__(self, numP):
        # number of particles
        self.numP = numP
        # time variable
        self.t = 0.0
        self.lifeTime = 5.0
        self.startPos = numpy.array([0.0, 0.0, 0.5])
        # load texture
        self.texid = glutils.loadTexture('star.png')
        # create shader
        self.program = glutils.loadShaders(strVS, strFS)
        glUseProgram(self.program)

        # set sampler
        texLoc = glGetUniformLocation(self.program, b"uTex")
        glUniform1i(texLoc, 0)

        # uniforms
        self.timeU = glGetUniformLocation(self.program, b"uTime")
        self.lifeTimeU = glGetUniformLocation(self.program, b"uLifeTime")
        self.pMatrixUniform = glGetUniformLocation(self.program, b'uPMatrix')
        self.mvMatrixUniform = glGetUniformLocation(self.program, b"uMVMatrix")
```

```python
        self.bMatrixU = glGetUniformLocation(self.program, b"bMatrix")
        self.colorU = glGetUniformLocation(self.program, b"uColor")
        self.samplerU = glGetUniformLocation(self.program, b"uSampler")
        self.posU = glGetUniformLocation(self.program, b"uPos")

        # attributes
        self.vertIndex = glGetAttribLocation(self.program, b"aVert")
        self.texIndex = glGetAttribLocation(self.program, b"aTexCoord")
        self.time0Index = glGetAttribLocation(self.program, b"aTime0")
        self.velIndex = glGetAttribLocation(self.program, b"aVel")

        # render flags
        self.enableBillboard = True
        self.disableDepthMask = True
        self.enableBlend = True

        # which texture to use
        self.useStarTexture = True
        # restart - first time
        self.restart(numP)

    # step
    def step(self):
        # increment time
        self.t += 0.01

    # restart particle system
    def restart(self, numP):
        # set number of particles
        self.numP = numP

        # time variables
        self.t = 0.0
        self.lifeTime = 5.0

        # color
        self.col0 = numpy.array([random.random(), random.random(),
                                 random.random(), 1.0])

        # create Vertex Arrays Object (VAO)
        self.vao = glGenVertexArrays(1)
        # bind VAO
        glBindVertexArray(self.vao)

        # create attribute arrays and vertex buffers:

        # vertices
        s = 0.2
        quadV = [
            -s,  s, 0.0,
            -s, -s, 0.0,
             s,  s, 0.0,
             s, -s, 0.0,
             s,  s, 0.0,
            -s, -s, 0.0
            ]
        vertexData = numpy.array(numP*quadV, numpy.float32)
        self.vertexBuffer = glGenBuffers(1)
        glBindBuffer(GL_ARRAY_BUFFER, self.vertexBuffer)
        glBufferData(GL_ARRAY_BUFFER, 4*len(vertexData), vertexData,
```

```python
                GL_STATIC_DRAW)

    # texture coordinates
    quadT = [
        0.0, 1.0,
        0.0, 0.0,
        1.0, 1.0,
        1.0, 0.0,
        1.0, 1.0,
        0.0, 0.0
        ]
    tcData = numpy.array(numP*quadT, numpy.float32)
    self.tcBuffer = glGenBuffers(1)
    glBindBuffer(GL_ARRAY_BUFFER, self.tcBuffer)
    glBufferData(GL_ARRAY_BUFFER, 4*len(tcData), tcData, GL_STATIC_DRAW)

    # time lags
    timeData = numpy.repeat(0.005*numpy.arange(numP, dtype=numpy.float32),
                4)
    self.timeBuffer = glGenBuffers(1)
    glBindBuffer(GL_ARRAY_BUFFER, self.timeBuffer)
    glBufferData(GL_ARRAY_BUFFER, 4*len(timeData), timeData,
                GL_STATIC_DRAW)
    # velocites
    velocities = []
    # cone angle
    coneAngle = math.radians(20.0)
    # set up particle velocities
    for i in range(numP):
        # inclination
        angleRatio = random.random()
        a = angleRatio*coneAngle
        # azimuth
        t = random.random()*(2.0*math.pi)
        # get veocity on sphere
        vx = math.sin(a)*math.cos(t)
        vy = math.sin(a)*math.sin(t)
        vz = math.cos(a)
        # speed decreases with angle
        speed = 15.0*(1.0 - angleRatio*angleRatio)
        # add a set of calculated velocities
        velocities += 6*[speed*vx, speed*vy, speed*vz]
    # set up velocity vertex buffer
    self.velBuffer = glGenBuffers(1)
    glBindBuffer(GL_ARRAY_BUFFER, self.velBuffer)
    velData = numpy.array(velocities, numpy.float32)
    glBufferData(GL_ARRAY_BUFFER, 4*len(velData), velData, GL_STATIC_DRAW)

    # enable arrays
    glEnableVertexAttribArray(self.vertIndex)
    glEnableVertexAttribArray(self.texIndex)
    glEnableVertexAttribArray(self.time0Index)
    glEnableVertexAttribArray(self.velIndex)

    # set buffers
    glBindBuffer(GL_ARRAY_BUFFER, self.vertexBuffer)
    glVertexAttribPointer(self.vertIndex, 3, GL_FLOAT, GL_FALSE, 0, None)

    glBindBuffer(GL_ARRAY_BUFFER, self.tcBuffer)
    glVertexAttribPointer(self.texIndex, 2, GL_FLOAT, GL_FALSE, 0, None)
```

```python
        glBindBuffer(GL_ARRAY_BUFFER, self.velBuffer)
        glVertexAttribPointer(self.velIndex, 3, GL_FLOAT, GL_FALSE, 0, None)

        glBindBuffer(GL_ARRAY_BUFFER, self.timeBuffer)
        glVertexAttribPointer(self.time0Index, 1, GL_FLOAT, GL_FALSE, 0, None)

        # unbind VAO
        glBindVertexArray(0)

    # render the particle system
    def render(self, pMatrix, mvMatrix, camera):
        # use shader
        glUseProgram(self.program)

        # set projection matrix
        glUniformMatrix4fv(self.pMatrixUniform, 1, GL_FALSE, pMatrix)

        # set modelview matrix
        glUniformMatrix4fv(self.mvMatrixUniform, 1, GL_FALSE, mvMatrix)
        # set up a billboard matrix to keep quad aligned to view direction
        if self.enableBillboard:
            N = camera.eye - camera.center
            N /= numpy.linalg.norm(N)
            U = camera.up
            U /= numpy.linalg.norm(U)
            R = numpy.cross(U, N)
            U2 = numpy.cross(N, R)
            bMatrix = numpy.array([R[0], U2[0], N[0], 0.0,
                                   R[1], U2[1], N[1], 0.0,
                                   R[2], U2[2], N[2], 0.0,
                                   0.0, 0.0, 0.0, 1.0], numpy.float32)
            glUniformMatrix4fv(self.bMatrixU, 1, GL_TRUE, bMatrix)
        else:
            # identity matrix
            bMatrix = numpy.array([1.0, 0.0, 0.0, 0.0,
                                   0.0, 1.0, 0.0, 0.0,
                                   0.0, 0.0, 1.0, 0.0,
                                   0.0, 0.0, 0.0, 1.0], numpy.float32)
            glUniformMatrix4fv(self.bMatrixU, 1, GL_FALSE, bMatrix)
        # set start position
        glUniform3fv(self.posU, 1, self.startPos)
        # set time
        glUniform1f(self.timeU, self.t)
        #set lifetime
        glUniform1f(self.lifeTimeU, self.lifeTime)
        # set color
        glUniform4fv(self.colorU, 1, self.col0)

        # enable texture
        glActiveTexture(GL_TEXTURE0)
        glBindTexture(GL_TEXTURE_2D, self.texid)
        glUniform1i(self.samplerU, 0)

        # turn depth mask off
        if self.disableDepthMask:
            glDepthMask(GL_FALSE)

        # enable blending
        if self.enableBlend:
```

```
        glBlendFunc(GL_SRC_ALPHA, GL_ONE)
        glEnable(GL_BLEND)

    # bind VAO
    glBindVertexArray(self.vao)
    # draw
    glDrawArrays(GL_TRIANGLES, 0, 6*self.numP)
    # unbind VAO
    glBindVertexArray(0)

    # disable blend
    if self.enableBlend:
        glDisable(GL_BLEND)

    # turn depth mask on
    if self.disableDepthMask:
        glDepthMask(GL_TRUE)

    # disable texture
    glBindTexture(GL_TEXTURE_2D, 0)
```

这是火花喷泉的所有代码，但我们还要画一个红色的盒子代表喷泉粒子系统的来源。

10.5 盒子代码

要让观众的注意力集中于喷泉，只要画一个没有任何灯光的红色立方体就行了。

```
import sys, random, math
import OpenGL
from OpenGL.GL import *
import numpy
import glutils

strVS = """
#version 330 core

in vec3 aVert;
uniform mat4 uMVMatrix;
uniform mat4 uPMatrix;
out vec4 vCol;

void main() {
    // apply transformations
    gl_Position = uPMatrix * uMVMatrix * vec4(aVert, 1.0);
    // set color
    vCol = vec4(0.8, 0.0, 0.0, 1.0);
}
"""

strFS = """
#version 330 core

in vec4 vCol;
out vec4 fragColor;

void main() {
```

```
        // use vertex color
        fragColor = vCol;
}
"""


class Box:
    def __init__(self, side):
        self.side = side

        # load shaders
        self.program = glutils.loadShaders(strVS, strFS)
        glUseProgram(self.program)

        s = side/2.0
        vertices = [
             -s,  s, -s,
             -s, -s, -s,
              s,  s, -s,
              s, -s, -s,
              s,  s, -s,
             -s, -s, -s,

             -s,  s,  s,
             -s, -s,  s,
              s,  s,  s,
              s, -s,  s,
              s,  s,  s,
             -s, -s,  s,

             -s, -s,  s,
             -s, -s, -s,
              s, -s,  s,
              s, -s, -s,
              s, -s,  s,
             -s, -s, -s,

             -s,  s,  s,
             -s,  s, -s,
              s,  s,  s,
              s,  s, -s,
              s,  s,  s,
             -s,  s, -s,

             -s, -s,  s,
             -s, -s, -s,
             -s,  s,  s,
             -s,  s, -s,
             -s,  s,  s,
             -s, -s, -s,

              s, -s,  s,
              s, -s, -s,
              s,  s,  s,
              s,  s, -s,
              s,  s,  s,
              s, -s, -s
             ]

        # set up vertex array object (VAO)
```

第 10 章 粒子系统

```python
        self.vao = glGenVertexArrays(1)
        glBindVertexArray(self.vao)
        # set up VBOs
        vertexData = numpy.array(vertices, numpy.float32)
        self.vertexBuffer = glGenBuffers(1)
        glBindBuffer(GL_ARRAY_BUFFER, self.vertexBuffer)
        glBufferData(GL_ARRAY_BUFFER, 4*len(vertexData), vertexData,
                     GL_STATIC_DRAW)
        #enable arrays
        self.vertIndex = glGetAttribLocation(self.program, "aVert")
        glEnableVertexAttribArray(self.vertIndex)
        # set buffers
        glBindBuffer(GL_ARRAY_BUFFER, self.vertexBuffer)
        glVertexAttribPointer(self.vertIndex, 3, GL_FLOAT, GL_FALSE, 0, None)
        # unbind VAO
        glBindVertexArray(0)

    def render(self, pMatrix, mvMatrix):

        # use shader
        glUseProgram(self.program)

        # set projection matrix
        glUniformMatrix4fv(glGetUniformLocation(self.program, 'uPMatrix'),
                           1, GL_FALSE, pMatrix)
        # set modelview matrix
        glUniformMatrix4fv(glGetUniformLocation(self.program, 'uMVMatrix'),
                           1, GL_FALSE, mvMatrix)
        # bind VAO
        glBindVertexArray(self.vao)
        # draw
        glDrawArrays(GL_TRIANGLES, 0, 36)
        # unbind VAO
        glBindVertexArray(0)
```

盒子的代码使用简单的顶点和片段着色器来绘制一个立方体。这里使用的概念与本章前面和第 9 章中讨论的相同。

10.6　主程序代码

该项目的主程序源文件是 psmain.py，它设置了 GLFW 窗口，处理了键盘事件，并创建粒子系统。如果想看到完整的程序代码，请跳到 10.7 节。

```
class PSMaker:
    """GLFW Rendering window class for Particle System"""
    def __init__(self):
❶       self.camera = Camera([15.0, 0.0, 2.5],
                             [0.0, 0.0, 2.5],
                             [0.0, 0.0, 1.0])
        self.aspect = 1.0
        self.numP = 300
        self.t = 0
        # flag to rotate camera view
        self.rotate = True

        # save current working directory
```

```
            cwd = os.getcwd()

            # initialize glfw; this changes cwd
❷           glfw.glfwInit()

            # restore cwd
            os.chdir(cwd)

            # version hints
            glfw.glfwWindowHint(glfw.GLFW_CONTEXT_VERSION_MAJOR, 3)
            glfw.glfwWindowHint(glfw.GLFW_CONTEXT_VERSION_MINOR, 3)
            glfw.glfwWindowHint(glfw.GLFW_OPENGL_FORWARD_COMPAT, GL_TRUE)
            glfw.glfwWindowHint(glfw.GLFW_OPENGL_PROFILE,
                                glfw.GLFW_OPENGL_CORE_PROFILE)

            # make a window
            self.width, self.height = 640, 480
            self.aspect = self.width/float(self.height)
            self.win = glfw.glfwCreateWindow(self.width, self.height,
                                             b"Particle System")

            # make context current
            glfw.glfwMakeContextCurrent(self.win)

            # initialize GL
            glViewport(0, 0, self.width, self.height)
            glEnable(GL_DEPTH_TEST)
            glClearColor(0.2, 0.2, 0.2,1.0)

            # set window callbacks
            glfw.glfwSetMouseButtonCallback(self.win, self.onMouseButton)
            glfw.glfwSetKeyCallback(self.win, self.onKeyboard)
            glfw.glfwSetWindowSizeCallback(self.win, self.onSize)

            # create 3D
❸           self.psys = ParticleSystem(self.numP)
❹           self.box = Box(1.0)

            # exit flag
❺           self.exitNow = False
```

类 **PSMaker** 创建粒子系统，处理 GLFW 窗口，并负责渲染喷泉和表示其源头的盒子。在❶行，创建一个 Camera 对象，它可以用于在 OpenGL 中设置观看参数。从❷行开始的代码块设置了 GLFW 窗口，在❸行，创建 ParticleSystem 对象，在❹行，创建 Box 对象。在❺行，是在主 GLFW 渲染循环中使用的退出标志，接下来你会看到。

10.6.1 每步更新这些粒子

通过在主程序循环中更新一个时间变量，我们创建了动画。主程序循环进而又更新顶点着色器中的时间变量。用这个新时间值，计算并渲染下一帧，从而更新这些粒子的位置。着色器计算粒子的新朝向，也让它们旋转。此外，着色器计算 alpha 值，它是时间变量的函数，将用于最后的渲染，实现火花逐渐淡出。

第 10 章 粒子系统 **167**

step()方法为每个时间步骤更新粒子系统。

```
    def step(self):
        # increment time
❶       self.t += 10
❷       self.psys.step()
        # rotate eye
        if self.rotate:
❸           self.camera.rotate()
        # restart every 5 seconds
        if not int(self.t) % 5000:
❹           self.psys.restart(self.numP)
```

在❶行，增加时间变量，它以毫秒为单位记录流逝的时间。在❷行，调用 ParticleSystem 的 step()方法，这样它可以自行更新。如果设置了标志，就在❸行旋转相机。每 5 秒钟（5000 毫秒），粒子系统在❹行重新启动。

10.6.2 键盘处理程序

现在，来看看 GLFW 窗口的键盘处理程序。

```
❶   def onKeyboard(self, win, key, scancode, action, mods):
        #print 'keyboard: ', win, key, scancode, action, mods
        if action == glfw.GLFW_PRESS:
            # ESC to quit
            if key == glfw.GLFW_KEY_ESCAPE:
                self.exitNow = True
            elif key == glfw.GLFW_KEY_R:
                self.rotate = not self.rotate
            elif key == glfw.GLFW_KEY_B:
                # toggle billboarding
                self.psys.enableBillboard = not self.psys.enableBillboard
            elif key == glfw.GLFW_KEY_D:
                # toggle depth mask
                self.psys.disableDepthMask = not self.psys.disableDepthMask
            elif key == glfw.GLFW_KEY_T:
                # toggle transparency
                self.psys.enableBlend = not self.psys.enableBlend
```

❶行的键盘处理程序，主要在关闭用于绘制粒子系统的各种渲染技巧时，让你很容易看到所发生的事情。这段代码让你切换旋转、公告板、深度遮掩和透明度。

10.6.3 管理主程序循环

使用 GLFW 时，必须管理自己的主程序循环。下面是这个程序中使用的循环：

```
    def run(self):
        # initializer timer
        glfw.SetTime(0)
        t = 0.0
❶       while not glfw.glfwWindowShouldClose(self.win) and not self.exitNow:
            # update every x seconds
❷           currT = glfw.glfwGetTime()
            if currT - t > 0.01:
                # update time
```

```
                t = currT

                # clear
                glClear(GL_COLOR_BUFFER_BIT | GL_DEPTH_BUFFER_BIT)

                # render
                pMatrix = glutils.perspective(100.0, self.aspect, 0.1, 100.0)
                # modelview matrix
                mvMatrix = glutils.lookAt(self.camera.eye, self.camera.center,
                                          self.camera.up)

                # draw nontransparent object first
❸               self.box.render(pMatrix, mvMatrix)

                # render
❹               self.psys.render(pMatrix, mvMatrix, self.camera)

                # step
❺               self.step()

                glfw.glfwSwapBuffers(self.win)
                # poll for and process events
                glfw.glfwPollEvents()
        # end
        glfw.glfwTerminate()
```

这段代码与第 9 章中使用的循环几乎是相同的。在❶行，如果 exit 标志已设置或 GLFW 窗口关闭，while 循环就退出。在❷行和后面一行，使用 GLFW 计时器，仅当时间流逝一定量（0.1 秒）时才进行渲染，从而控制渲染的帧速率。在❸行绘制盒子，在❹行绘制粒子系统（渲染的顺序很重要：透明物体总是最后绘制，这样它们能够与场景中不透明的物体正确地混合和深度缓冲）。在❺行，针对当前时间步骤更新粒子系统。

10.7 完整主程序代码

下面是 psmain.py 的完整代码。也可以在 https://github.com/electronut/pp/tree/master/particle-system/ 找到这段代码。

```
import sys, os, math, numpy
import OpenGL
from OpenGL.GL import *
import numpy
from ps import ParticleSystem, Camera
from box import Box
import glutils
import glfw

class PSMaker:
    """GLFW Rendering window class for Particle System"""
    def __init__(self):
        self.camera = Camera([15.0, 0.0, 2.5],
                             [0.0, 0.0, 2.5],
                             [0.0, 0.0, 1.0])
```

第 10 章　粒子系统　169

```python
        self.aspect = 1.0
        self.numP = 300
        self.t = 0
        # flag to rotate camera view
        self.rotate = True

        # save current working directory
        cwd = os.getcwd()

        # initialize glfw; this changes cwd
        glfw.glfwInit()

        # restore cwd
        os.chdir(cwd)

        # version hints
        glfw.glfwWindowHint(glfw.GLFW_CONTEXT_VERSION_MAJOR, 3)
        glfw.glfwWindowHint(glfw.GLFW_CONTEXT_VERSION_MINOR, 3)
        glfw.glfwWindowHint(glfw.GLFW_OPENGL_FORWARD_COMPAT, GL_TRUE)
        glfw.glfwWindowHint(glfw.GLFW_OPENGL_PROFILE,
                            glfw.GLFW_OPENGL_CORE_PROFILE)

        # make a window
        self.width, self.height = 640, 480
        self.aspect = self.width/float(self.height)
        self.win = glfw.glfwCreateWindow(self.width, self.height,
                                         b"Particle System")
        # make context current
        glfw.glfwMakeContextCurrent(self.win)

        # initialize GL
        glViewport(0, 0, self.width, self.height)
        glEnable(GL_DEPTH_TEST)
        glClearColor(0.2, 0.2, 0.2,1.0)

        # set window callbacks
        glfw.glfwSetMouseButtonCallback(self.win, self.onMouseButton)
        glfw.glfwSetKeyCallback(self.win, self.onKeyboard)
        glfw.glfwSetWindowSizeCallback(self.win, self.onSize)

        # create 3D
        self.psys = ParticleSystem(self.numP)
        self.box = Box(1.0)

        # exit flag
        self.exitNow = False

    def onMouseButton(self, win, button, action, mods):
        #print 'mouse button: ', win, button, action, mods
        pass

    def onKeyboard(self, win, key, scancode, action, mods):
        #print 'keyboard: ', win, key, scancode, action, mods
        if action == glfw.GLFW_PRESS:
            # ESC to quit
            if key == glfw.GLFW_KEY_ESCAPE:
                self.exitNow = True
            elif key == glfw.GLFW_KEY_R:
                self.rotate = not self.rotate
```

```python
            elif key == glfw.GLFW_KEY_B:
                # toggle billboarding
                self.psys.enableBillboard = not self.psys.enableBillboard
            elif key == glfw.GLFW_KEY_D:
                # toggle depth mask
                self.psys.disableDepthMask = not self.psys.disableDepthMask
            elif key == glfw.GLFW_KEY_T:
                # toggle transparency
                self.psys.enableBlend = not self.psys.enableBlend

    def onSize(self, win, width, height):
        #print 'onsize: ', win, width, height
        self.width = width
        self.height = height
        self.aspect = width/float(height)
        glViewport(0, 0, self.width, self.height)

    def step(self):
        # increment time
        self.t += 10
        self.psys.step()
        # rotate eye
        if self.rotate:
            self.camera.rotate()
        # restart every 5 seconds
        if not int(self.t) % 5000:
            self.psys.restart(self.numP)

    def run(self):
        # initializer timer
        glfw.glfwSetTime(0)
        t = 0.0
        while not glfw.glfwWindowShouldClose(self.win) and not self.exitNow:
            # update every x seconds
            currT = glfw.glfwGetTime()
            if currT - t > 0.01:
                # update time
                t = currT

                # clear
                glClear(GL_COLOR_BUFFER_BIT | GL_DEPTH_BUFFER_BIT)

                # render
                pMatrix = glutils.perspective(100.0, self.aspect, 0.1, 100.0)
                # modelview matrix
                mvMatrix = glutils.lookAt(self.camera.eye, self.camera.center,
                                          self.camera.up)

                # draw nontransparent object first
                self.box.render(pMatrix, mvMatrix)

                # render
                self.psys.render(pMatrix, mvMatrix, self.camera)

                # step
                self.step()

                glfw.glfwSwapBuffers(self.win)
                # poll for and process events
                glfw.glfwPollEvents()
```

```
        # end
        glfw.glfwTerminate()
# main() function
def main():
    # use sys.argv if needed
    print('starting particle system...')
    prog = PSMaker()
    prog.run()

# call main
if __name__ == '__main__':
    main()
```

10.8 运行程序

要运行该项目,请输入以下命令:

```
$ python3 psmain.py
```

本章开始的图 10-1 展示了输出。

10.9 小结

本章用 Python 和 OpenGL 创建了喷泉粒子系统。我们学习了创建粒子系统的数学模型,建立了着色器程序,并利用了一些 OpenGL 技巧,以逼真的方式渲染粒子。

10.10 实验

下面有一些想法,可以用更多的方式来试验粒子系统动画。

1. 如果每个粒子的喷射没有时间延迟,看看情况如何。

2. 让喷泉中的粒子在喷射时逐渐变大(提示:缩放顶点着色器的四顶点)。

3. 这些代码让每个粒子遵循完美的抛物线,因为它从喷泉喷出。请为粒子的路径增加一些随机性(提示:研究 GLSL 的 noise() 方法,你可以在顶点着色器中使用)。

4. 在喷泉中的粒子遵循抛物线轨迹,上升,然后由于重力作用下落。你能让它们跌落到地板时反弹吗?可能需要增加粒子的寿命来实现(提示:地板位于 z = 0.0 处。在顶点着色器中,当粒子接近地面时,反转速度的 z 分量)。

第 11 章
体渲染

MRI 和 CT 扫描这样的诊断过程会创建体数据,它由一组二维图像来表示通过三维物体的截面。体绘制是一种计算机图形技术,利用这种类型的体数据来构造三维图像。虽然体绘制通常用于分析医学扫描,但也可以用于地质、考古和分子生物学等学科,生成三维科学可视化效果。

由 MRI 和 CT 扫描记录的数据通常采取 $N_x \times N_y \times N_z$ 的三维网格形式,或 N_z 个二维"切片",其中每个切片是大小为 $N_x \times N_y$ 的图像。体绘制算法采用某种类型的透明度,来展示采集切片数据,并采用各种技术来强调被渲染物体中关注的部分。

本项目将探讨名为"体光线投射(volume ray casting)"的体绘制算法,它充分利用图形处理单元(GPU),用 OpenGL 着色语言(GLSL)的着色器来进行计算。代码针对屏幕上的每个像素执行,并利用了 GPU,其目的是有效地进行并行计算。利用二维图像文件夹中包含的三维数据集切片,用体光线投射算法,构建体渲染图像。我们还会实现一种方法,显示在 x、y、z 方向上的二维切片数据,以便用户可以通过箭头键来滚动切片。键盘命令让用户在三维渲染和二维切片之间切换。

下面是本项目涉及的一些主题：
- 用 GLSL 进行 GPU 计算；
- 创建顶点和片段着色器；
- 表示三维立体数据，并使用体光线投射算法；
- 用 numpy 数组表示三维变换矩阵。

11.1 工作原理

有许多方式来渲染三维数据集。本项目将使用体光线投射的方法，它是基于图像的渲染技术，从二维切片逐个像素地生成最终图像的。与此形成对比的是，典型的三维渲染方法是基于对象的：它们开始用三维对象表示，然后应用变换来生成投射的二维图像中的像素。

在本项目用到的体光线投射方法中，对于输出图像中的每个像素，一束光线射入离散的三维体数据集，其通常表示为一个长方体。随着光线穿过物体，数据以一定的间隔采样，样本被组合（或"合成"）来计算最终图像的颜色值或强度（你可以认为这个过程类似于堆叠一些透明胶片，将它们放在强光下，看到所有的胶片组合的效果）。

体光线投射渲染的实现通常使用一些技术，如采用梯度改进最终渲染的外观，用过滤来分离三维特征，用空间优化技术以提高速度，但我们只实现基本的光线投射算法，通过 x 射线投射合成最终图像（我的实现，主要基于 2003 年 Kruger 和 Westermann 在该主题上的开创性论文[1]）。

11.1.1 数据格式

本项目将使用来自斯坦福体数据档案（Stanford Volume Data Archive）的三维扫描医学数据[2]。该档案提供了一些极好的三维医学数据集的 TIFF 图像（包括 CT 和 MRI），每张图像表示该物体的每个二维横截面。我们将读入文件夹中的这些图像，作为 OpenGL 三维纹理。这有点像堆叠一系列二维图像，形成一个长方体，如图 11-1 所示。

回顾第 9 章，OpenGL 中的二维纹理是由一个二维坐标（s，t）定位的。类似地，三维纹理用三维纹理坐标定位，形如（s，t，p）。如你所见，将体数据保存为三维纹理，可以让你快速访问数据，并提供光线投射方案所需的插值。

---

[1] J. Kruger and R. Westermann, "Acceleration Techniques for GPU-based Volume Rendering," IEEE Visualization, 2003.
[2] http://graphics.stanford.edu/data/voldata/

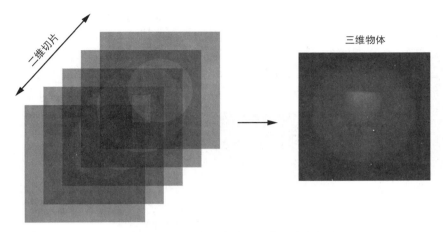

图 11-1 从二维切片建立三维立体数据

11.1.2 生成光线

该项目的目标是生成三维体数据的透视投影，如图 11-2 所示。

图 11-2 展示了 OpenGL 视锥，这在第 9 章讨论过。具体来说，它展示了来自眼睛的光线如何从近裁剪平面进入此锥体，通过立方状的物体（其中包含体数据），并从后面的远裁剪平面出去。

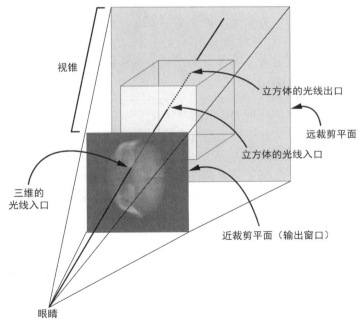

图 11-2 三维体数据透视投影

为了实现光线投射，需要生成进入物体的光线。对于图 11-2 所示的输出窗口中的每个像素，生成一个向量 R，它进入我们作为单位立方体的物体中（我称之为彩色立方体），该立方体在坐标（0，0，0）和（1，1，1）之间。这个立方体中的每个点的颜色，其 RGB 值等于它的三维坐标。原点颜色为（0，0，0），即黑色。(1，0，0) 角是红色，与原点相对的顶点颜色为（1，1，1），即白色。图 11-3 展示了这个立方体。

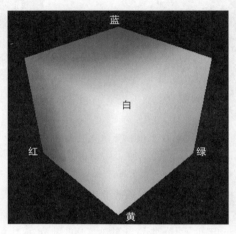

图 11-3　彩色立方体

注意　　在 OpenGL 中，颜色可以表示为一串 8 位无符号值(r, g, b)，其中 r、g 和 b 的范围是[0, 255]。它也可以表示为 32 位的浮点值(r, g, b)，其中 r、g 和 b 的范围是[0.0, 1.0]。这些表示是等效的。例如，前一种的红色（255，0，0）等同于后一种的（1.0，0.0，0.0）。

要绘制立方体，先用 OpenGL 图元 GL_TRIANGLES 绘制它的 6 个面。然后确定每个顶点的颜色，并在多边形光栅化、生成顶点之间的颜色时，利用 OpenGL 提供的插值。例如，如图 11-4（A）展示了立方体的三个正面。图 11-4（B）将 OpenGL 设置为剔除正面，绘制了立方体的背面。

如果从图 11-4（B）中减去图 11-4（A）中的颜色，即从(r, g, b)back 减去(r, g, b)front，实际上计算了一组向量，从立方体的正面指向背面，因为该立方体上每点的颜色(r, g, b)与三维坐标相同。图 11-4（C）展示了结果（为了展示，负值已经翻转为正，因为负数不能直接显示为颜色）。读取一个像素的颜色值(r, g, b)，如图 11-4（C）所示，得到（rx, ry, rz）坐标，即在这一点穿透物体的光线。

有了投射光线后，将它们渲染成一张图或二维纹理，稍后与 OpenGL 的帧缓冲对象（FBO）功能一起使用。产生这个纹理之后，可以在实现光线投射算法的着色器内访问它。

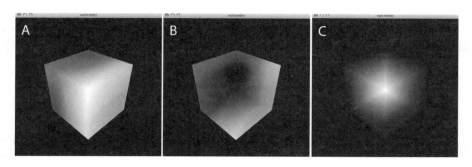

图 11-4　用于计算光线的颜色立方体

GPU 中的光线投射

为了实现光线投射算法，首先将彩色立方体的背面绘制到 FBO。接着，将正面绘制在屏幕上。光线投射算法的主要部分发生在片段着色器中，进行这种二次渲染，针对输出的每个像素进行。光线的计算是用颜色立方体的背面颜色减去传入片段的正面颜色，背面颜色是从纹理中读取的。然后将计算好的光线累加，并利用着色器中可用的三维体纹理数据，计算最终的像素值。

> **背面剔除**
>
> 在 OpenGL 中，如果要画四边形这样的图元，指定顶点的顺序很重要。默认情况下，OpenGL 假定用 GL_CCW（逆时针）的顺序。对于逆时针顺序的图元，法矢量将指向多边形的"外面"。如果试图绘制封闭区域的几何图形，如立方体，这就有关系了。为了优化渲染，你不用绘制看不见的面，只要打开 OpenGL 的"背面剔除（back-face culling）"，计算观看方向向量与多边形法向量的点积。对于朝向背面的多边形，点积是负的，这些面可以丢弃。OpenGL 也提供了一种简单的方法来做相反的事，即正面剔除，这就是我们在算法要用到的。

显示二维切片

除了三维渲染，也可以显示数据的二维切片，做法是从三维数据中提取垂直于 x 轴、y 轴或 z 轴的二维横截面，并将它作为一个四边形的纹理。因为我们将物体保存为三维纹理，所以很容易通过指定纹理坐标（s, t, p）得到所需的数据。OpenGL 内置的纹理插值可以给出三维纹理内任意位置的纹理值。

11.1.3　显示 OpenGL 窗口

像其他 OpenGL 项目一样，本项目用 GLFW 库来显示 OpenGL 窗口。我们用处理程序来绘制，响应调整窗口大小和键盘事件。我们将利用键盘事件在体渲染和切片渲染之间切换，以及实现穿过三维数据的旋转和切片。

11.2 所需模块

我们用 PyOpenGL 进行渲染，它是流行的 OpenGL 的 Python 绑定。我们还用 numpy 数组来表示三维坐标和变换矩阵。

11.3 项目代码概述

我们首先从体数据生成三维纹理，数据是从文件读入的。接下来，我们分析颜色立方体技术，用于生成从眼睛指向物体的光线，这是实现体光线投射算法的关键概念。我们将分析如何定义立方体的几体形状，如何绘制该立方体的背面和正面。然后，我们将探索体积光线投射算法，以及相关的顶点和片段着色器。最后，我们将学习如何实现体数据的二维切片。

该项目有 7 个 Python 文件：

glutils.py 包含针对 OpenGL 着色器、转换和其他相关工具方法；

makedata.py 包含创建测试体数据的工具方法；

raycast.py 实现了 RayCastRender 类，用于光线投射；

raycube.py 实现了 RayCastRender 中使用的 RayCube 类；

slicerender.py 实现了 SliceRender 类，用于体数据的二维切片；

volreader.py 包含一个工具方法，读取体数据作为 OpenGL 的三维纹理；

volrender.py 包含用于创建 GLFW 窗口和渲染器的主要方法。

除了两个文件之外，本章将介绍所有这些文件。文件 makedata.py 和本章中的其他项目文件一起放在 https://github.com/electronut/pp/tree/master/volrender/。文件 glutils.py 可以从 https://github.com/electronut/pp/tree/master/common/下载。

11.4 生成三维纹理

第一步是从包含图像的文件夹中读取体数据，如下面代码所示。要查看 volreader.py 的完整代码，请直接跳到 11.5 节。

```
def loadVolume(dirName):
    """read volume from directory as a 3D texture"""
    # list images in directory
❶   files = sorted(os.listdir(dirName))
    print('loading images from: %s' % dirName)
    imgDataList = []
    count = 0
    width, height = 0, 0
    for file in files:
❷       file_path = os.path.abspath(os.path.join(dirName, file))
        try:
```

```
                # read image
❸               img = Image.open(file_path)
                imgData = np.array(img.getdata(), np.uint8)

                # check if all images are of the same size
❹               if count is 0:
                    width, height = img.size[0], img.size[1]
                    imgDataList.append(imgData)
                else:
❺                   if (width, height) == (img.size[0], img.size[1]):
                        imgDataList.append(imgData)
                    else:
                        print('mismatch')
                        raise RunTimeError("image size mismatch")
                count += 1
                #print img.size
            except:
                # skip
                print('Invalid image: %s' % file_path)

        # load image data into single array
        depth = count
❻       data = np.concatenate(imgDataList)
        print('volume data dims: %d %d %d' % (width, height, depth))

        # load data into 3D texture
❼       texture = glGenTextures(1)
        glPixelStorei(GL_UNPACK_ALIGNMENT, 1)
        glBindTexture(GL_TEXTURE_3D, texture)
        glTexParameterf(GL_TEXTURE_3D, GL_TEXTURE_WRAP_S, GL_CLAMP_TO_EDGE)
        glTexParameterf(GL_TEXTURE_3D, GL_TEXTURE_WRAP_T, GL_CLAMP_TO_EDGE)
        glTexParameterf(GL_TEXTURE_3D, GL_TEXTURE_WRAP_R, GL_CLAMP_TO_EDGE)
        glTexParameterf(GL_TEXTURE_3D, GL_TEXTURE_MAG_FILTER, GL_LINEAR)
        glTexParameterf(GL_TEXTURE_3D, GL_TEXTURE_MIN_FILTER, GL_LINEAR)
❽       glTexImage3D(GL_TEXTURE_3D, 0, GL_RED,
                     width, height, depth, 0,
                     GL_RED, GL_UNSIGNED_BYTE, data)
        # return texture
❾       return (texture, width, height, depth)
```

loadVolume()方法先用 os 模块中的 listdir()方法，列出了给定目录中的文件❶。然后加载图像文件本身。在❷行，利用 os.path.abspath()和 os.path.join()，将文件名添加到目录，从而不必处理相对文件路径和操作系统（OS）特定的路径约定（Python 代码中经常可以看到这个有用的习惯用法，用于遍历文件和目录）。

在❸行，利用 Python 图像库（PIL）中的 Image 类，将图像加载到 8 位的 numpy 数组。如果指定的文件不是图像或图像加载失败，则抛出一个异常，我们捕获它并打印错误。

因为要将这些图像加载成为三维纹理，所以需要确保它们都具有相同的尺寸（宽×高），这在❹和❺行确认。保存第一张图像的尺寸，并与新传入的图像进行比较。所有图像加载为单独的数组后，利用 numpy 的 concatenate()方法，将这些数组连接起来，创建包含三维数据的最终数组❻。

在❼行和接下来几行中,创建了 OpenGL 纹理,并针对过滤和拆包设置了参数。然后,在❽行,将三维数据数组加载为 OpenGL 纹理。这里使用的格式是 GL_RED,数据格式是 GL_UNSIGNED_BYTE,因为在数据中,我们只有一个 8 位值与每个像素相关联。

最后,在❾行,返回了 OpenGL 纹理 ID 和三维纹理的尺寸。

11.5 完整的三维纹理代码

下面是完整的代码清单。可以在 https:// github.com/electronut/pp/tree/master/volrender/找到 volreader.py 文件。

```python
import os
import numpy as np
from PIL import Image

import OpenGL
from OpenGL.GL import *

from scipy import misc

def loadVolume(dirName):
    """read volume from directory as a 3D texture"""
    # list images in directory
    files = sorted(os.listdir(dirName))
    print('loading images from: %s' % dirName)
    imgDataList = []
    count = 0
    width, height = 0, 0
    for file in files:
        file_path = os.path.abspath(os.path.join(dirName, file))
        try:
            # read image
            img = Image.open(file_path)
            imgData = np.array(img.getdata(), np.uint8)

            # check if all are of the same size
            if count is 0:
                width, height = img.size[0], img.size[1]
                imgDataList.append(imgData)
            else:
                if (width, height) == (img.size[0], img.size[1]):
                    imgDataList.append(imgData)
                else:
                    print('mismatch')
                    raise RunTimeError("image size mismatch")
            count += 1
            #print img.size
        except:
            # skip
            print('Invalid image: %s' % file_path)

    # load image data into single array
    depth = count
    data = np.concatenate(imgDataList)
```

```
        print('volume data dims: %d %d %d' % (width, height, depth))

        # load data into 3D texture
        texture = glGenTextures(1)
        glPixelStorei(GL_UNPACK_ALIGNMENT, 1)
        glBindTexture(GL_TEXTURE_3D, texture)
        glTexParameterf(GL_TEXTURE_3D, GL_TEXTURE_WRAP_S, GL_CLAMP_TO_EDGE)
        glTexParameterf(GL_TEXTURE_3D, GL_TEXTURE_WRAP_T, GL_CLAMP_TO_EDGE)
        glTexParameterf(GL_TEXTURE_3D, GL_TEXTURE_WRAP_R, GL_CLAMP_TO_EDGE)
        glTexParameterf(GL_TEXTURE_3D, GL_TEXTURE_MAG_FILTER, GL_LINEAR)
        glTexParameterf(GL_TEXTURE_3D, GL_TEXTURE_MIN_FILTER, GL_LINEAR)
        glTexImage3D(GL_TEXTURE_3D, 0, GL_RED,
                    width, height, depth, 0,
                    GL_RED, GL_UNSIGNED_BYTE, data)
        #return texture
        return (texture, width, height, depth)


# load texture
def loadTexture(filename):
    img = Image.open(filename)
    img_data = np.array(list(img.getdata()), 'B')
    texture = glGenTextures(1)
    glPixelStorei(GL_UNPACK_ALIGNMENT,1)
    glBindTexture(GL_TEXTURE_2D, texture)
    glTexParameterf(GL_TEXTURE_2D, GL_TEXTURE_WRAP_S, GL_CLAMP_TO_EDGE)
    glTexParameterf(GL_TEXTURE_2D, GL_TEXTURE_WRAP_T, GL_CLAMP_TO_EDGE)
    glTexParameterf(GL_TEXTURE_2D, GL_TEXTURE_MAG_FILTER, GL_LINEAR)
    glTexParameterf(GL_TEXTURE_2D, GL_TEXTURE_MIN_FILTER, GL_LINEAR)
    glTexImage2D(GL_TEXTURE_2D, 0, GL_RGBA, img.size[0], img.size[1],
                    0, GL_RGBA, GL_UNSIGNED_BYTE, img_data)
    return texture
```

11.6 生成光线

生成光线的代码封装在 RayCube 类中。该类负责绘制彩色立方体，它有一些方法将立方体的背面绘制到 FBO 或纹理中，并将立方体的正面绘制到屏幕上。要查看完整的 raycube.py 代码，请直接跳到 11.7 节。

首先，定义这个类使用的着色器：

```
❶ strVS = """
#version 330 core

layout(location = 1) in vec3 cubePos;
layout(location = 2) in vec3 cubeCol;

uniform mat4 uMVMatrix;
uniform mat4 uPMatrix;
out vec4 vColor;

void main()
{
    // set back-face color
    vColor = vec4(cubeCol.rgb, 1.0);
```

```
            // transformed position
            vec4 newPos = vec4(cubePos.xyz, 1.0);

            // set position
            gl_Position = uPMatrix * uMVMatrix * newPos;
        }
        """
❷  strFS = """
        #version 330 core

        in vec4 vColor;
        out vec4 fragColor;

        void main()
        {
            fragColor = vColor;
        }
        """
```

在❶行，定义了 RayCube 类使用的顶点着色器。该着色器有两个输入属性：cubePos 和 cubeCol，它们分别用于访问顶点的位置和颜色的值。模型视图和投影矩阵分别通过 uniform 变量 uMVMatrix 和 pMatrix 传入。vColor 变量声明为输出，因为它需要传入片段着色器，在那里进行插值。在❷行实现的片段着色器，将片段颜色设置为传入的 vColor（插值后的）值，vColor 在顶点着色器中设置。

11.6.1 定义颜色立方体的几何形状

现在，看看颜色立方体的几何形状，它在 RayCube 类中定义：

```
            # cube vertices
❶           vertices = numpy.array([
                    0.0, 0.0, 0.0,
                    1.0, 0.0, 0.0,
                    1.0, 1.0, 0.0,
                    0.0, 1.0, 0.0,
                    0.0, 0.0, 1.0,
                    1.0, 0.0, 1.0,
                    1.0, 1.0, 1.0,
                    0.0, 1.0, 1.0
                    ], numpy.float32)

            # cube colors
❷           colors = numpy.array([
                    0.0, 0.0, 0.0,
                    1.0, 0.0, 0.0,
                    1.0, 1.0, 0.0,
                    0.0, 1.0, 0.0,
                    0.0, 0.0, 1.0,
                    1.0, 0.0, 1.0,
                    1.0, 1.0, 1.0,
                    0.0, 1.0, 1.0
                    ], numpy.float32)

            # individual triangles
❸           indices = numpy.array([
```

```
            4, 5, 7,
            7, 5, 6,
            5, 1, 6,
            6, 1, 2,
            1, 0, 2,
            2, 0, 3,
            0, 4, 3,
            3, 4, 7,
            6, 2, 7,
            7, 2, 3,
            4, 0, 5,
            5, 0, 1
            ], numpy.int16)
```

在 RayCube 的构造函数中，汇集了着色器，并创建了 program 对象。在❶行定义了立方体的几何形状，❷行定义了颜色。

颜色立方体有 6 个面，每个面可以绘制成两个三角形，总共有 6×6（36）个顶点。但是，我们不是指定所有 36 个顶点，而是指定立方体的 8 个顶点，然后用一个 indices 数组定义一些三角形，如❸行和图 11-5 所示。

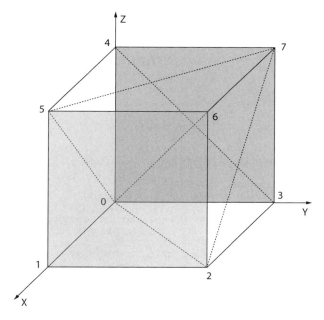

图 11-5　利用索引，一个立方体可以表示为三角形的集合，每个面由两个三角形组成

接下来，需要将顶点信息放入缓冲区。

```
            # set up vertex array object (VAO)
            self.vao = glGenVertexArrays(1)
            glBindVertexArray(self.vao)

            # vertex buffer
            self.vertexBuffer = glGenBuffers(1)
```

```
        glBindBuffer(GL_ARRAY_BUFFER, self.vertexBuffer)
        glBufferData(GL_ARRAY_BUFFER, 4*len(vertices), vertices, GL_STATIC_DRAW)

        # vertex buffer – cube vertex colors
        self.colorBuffer = glGenBuffers(1)
        glBindBuffer(GL_ARRAY_BUFFER, self.colorBuffer)
        glBufferData(GL_ARRAY_BUFFER, 4*len(colors), colors, GL_STATIC_DRAW)

        # index buffer
        self.indexBuffer = glGenBuffers(1)
❶       glBindBuffer(GL_ELEMENT_ARRAY_BUFFER, self.indexBuffer);
        glBufferData(GL_ELEMENT_ARRAY_BUFFER, 2*len(indices), indices,
                     GL_STATIC_DRAW)
```

像以往的项目一样，我们可以创建并绑定到一个顶点数组对象（VAO），然后定义它管理的缓冲区。这里有一个区别：在❶行，indices 数组设定为 GL_ELEMENT_ARRAY_BUFFER，这意味着它的缓冲器中的元素将用来索引和访问颜色和顶点缓冲器中的数据。

11.6.2 创建帧缓冲区对象

现在，让我们跳到创建帧缓冲区对象的方法，在该方法中将进行渲染。

```
    def initFBO(self):
        # create frame buffer object
        self.fboHandle = glGenFramebuffers(1)
        # create texture
        self.texHandle = glGenTextures(1)
        # create depth buffer
        self.depthHandle = glGenRenderbuffers(1)

        # bind
        glBindFramebuffer(GL_FRAMEBUFFER, self.fboHandle)

        glActiveTexture(GL_TEXTURE0)
        glBindTexture(GL_TEXTURE_2D, self.texHandle)

        # set parameters to draw the image at different sizes
❶       glTexParameteri(GL_TEXTURE_2D, GL_TEXTURE_MIN_FILTER, GL_LINEAR)
        glTexParameteri(GL_TEXTURE_2D, GL_TEXTURE_MAG_FILTER, GL_LINEAR)
        glTexParameteri(GL_TEXTURE_2D, GL_TEXTURE_WRAP_S, GL_CLAMP_TO_EDGE)
        glTexParameteri(GL_TEXTURE_2D, GL_TEXTURE_WRAP_T, GL_CLAMP_TO_EDGE)
        # set up texture
        glTexImage2D(GL_TEXTURE_2D, 0, GL_RGBA, self.width, self.height,
                     0, GL_RGBA, GL_UNSIGNED_BYTE, None)

        # bind texture to FBO
❷       glFramebufferTexture2D(GL_FRAMEBUFFER, GL_COLOR_ATTACHMENT0,
                               GL_TEXTURE_2D, self.texHandle, 0)

        # bind
❸       glBindRenderbuffer(GL_RENDERBUFFER, self.depthHandle)
        glRenderbufferStorage(GL_RENDERBUFFER, GL_DEPTH_COMPONENT24,
                              self.width, self.height)

        # bind depth buffer to FBO
        glFramebufferRenderbuffer(GL_FRAMEBUFFER, GL_DEPTH_ATTACHMENT,
```

```
                              GL_RENDERBUFFER, self.depthHandle)
            # check status
❹           status = glCheckFramebufferStatus(GL_FRAMEBUFFER)
            if status == GL_FRAMEBUFFER_COMPLETE:
                pass
                #print "fbo %d complete" % self.fboHandle
            elif status == GL_FRAMEBUFFER_UNSUPPORTED:
                print "fbo %d unsupported" % self.fboHandle
            else:
                print "fbo %d Error" % self.fboHandle
```

这里，我们创建帧缓冲区对象、二维纹理和渲染缓冲区对象。然后在❶行，建立了纹理参数。在❷行，纹理绑定到帧缓冲区对象。在❸行和随后的几行中，渲染缓冲区设置了一个24位的深度缓冲区，并连接到帧缓冲区对象。在❹行，检查帧缓冲区的状态，如果出现错误则打印状态信息。现在，只要帧缓冲区和渲染缓冲区绑定正确，所有的渲染将进入纹理。

11.6.3 渲染立方体的背面

下面的代码渲染了颜色立方体的背面：

```
        def renderBackFace(self, pMatrix, mvMatrix):
            """renders back-face of ray-cube to a texture and returns it"""
            # render to FBO
❶           glBindFramebuffer(GL_FRAMEBUFFER, self.fboHandle)
            # set active texture
            glActiveTexture(GL_TEXTURE0)
            # bind to FBO texture
            glBindTexture(GL_TEXTURE_2D, self.texHandle)

            # render cube with face culling enabled
❷           self.renderCube(pMatrix, mvMatrix, self.program, True)

            # unbind texture
❸           glBindTexture(GL_TEXTURE_2D, 0)
            glBindFramebuffer(GL_FRAMEBUFFER, 0)
            glBindRenderbuffer(GL_RENDERBUFFER, 0)

            # return texture ID
❹           return self.texHandle
```

在❶行，绑定FBO，设置活动纹理单元，并绑定到纹理处理程序，这样就可以渲染到FBO。在❷行，调用RayCube的renderCube()方法，以一个面剔除标志为参数，以便用同样的代码来绘制立方体的正面或背面。该标志设置为true，让背面出现在FBO纹理中。在❸行，进行必要的调用，解除FBO绑定，这样其他渲染代码将不受影响。在❹行返回FBO纹理ID，用于算法的下一阶段。

11.6.4 渲染立方体的正面

以下代码在光线投射算法的第二遍渲染时，绘制颜色立方体的正面。它就调用了11.6.3小节中讨论的renderCube()方法，只是面剔除标志设置为False。

```
def renderFrontFace(self, pMatrix, mvMatrix, program):
    """render front-face of ray-cube"""
    # no face culling
    self.renderCube(pMatrix, mvMatrix, program, False)
```

11.6.5 渲染整个立方体

现在看一下 renderCube()方法,它绘制前面讨论过的彩色立方体:

```
def renderCube(self, pMatrix, mvMatrix, program, cullFace):
    """renderCube uses face culling if flag set"""

    glClear(GL_COLOR_BUFFER_BIT | GL_DEPTH_BUFFER_BIT)

    # set shader program
    glUseProgram(program)

    # set projection matrix
    glUniformMatrix4fv(glGetUniformLocation(program, b'uPMatrix'),
                       1, GL_FALSE, pMatrix)

    # set modelview matrix
    glUniformMatrix4fv(glGetUniformLocation(program, b'uMVMatrix'),
                       1, GL_FALSE, mvMatrix)

    # enable face culling
    glDisable(GL_CULL_FACE)
```
❶
```
    if cullFace:
        glFrontFace(GL_CCW)
        glCullFace(GL_FRONT)
        glEnable(GL_CULL_FACE)

    # bind VAO
    glBindVertexArray(self.vao)

    # animated slice
```
❷
```
    glDrawElements(GL_TRIANGLES, self.nIndices, GL_UNSIGNED_SHORT, None)

    # unbind VAO
    glBindVertexArray(0)

    # reset cull face
    if cullFace:
        # disable face culling
        glDisable(GL_CULL_FACE)
```

在这个列表中可以看到,我们清除了颜色和深度缓冲区,然后选择着色器程序,设置了变换矩阵。在❶行,设置了一个标志,控制面剔除,这就决定了绘制立方体的正面或背面。而且,我们使用了 glDrawElements()❷,因为要用一个索引数组来渲染立方体,而不是一个顶点数组。

11.6.6 调整大小处理程序

由于 FBO 是针对特定窗口大小创建的，所以窗口大小发生变化时，需要重新创建它。要做到这一点，可以为 RayCube 类创建调整大小的处理程序，如下所示：

```
❶   def reshape(self, width, height):
        self.width = width
        self.height = height
        self.aspect = width/float(height)
        # re-create FBO
        self.clearFBO()
        self.initFBO()
```

调整 OpenGL 窗口的大小时，调用了 reshape()函数❶。

11.7 完整的光线生成代码

以下是完整的代码清单。也可以 https://github.com/electronut/pp/tree/master/volrender/找到 raycube.py 文件。

```
import OpenGL
from OpenGL.GL import *
from OpenGL.GL.shaders import *

import numpy, math, sys
import volreader, glutils

strVS = """
#version 330 core

layout(location = 1) in vec3 cubePos;
layout(location = 2) in vec3 cubeCol;

uniform mat4 uMVMatrix;
uniform mat4 uPMatrix;
out vec4 vColor;

void main()
{
    // set back face color
    vColor = vec4(cubeCol.rgb, 1.0);

    // transformed position
    vec4 newPos = vec4(cubePos.xyz, 1.0);

    // set position
    gl_Position = uPMatrix * uMVMatrix * newPos;

}
"""
strFS = """
#version 330 core

in vec4 vColor;
```

```
out vec4 fragColor;

void main()
{
    fragColor = vColor;
}
"""

class RayCube:
    """class used to generate rays used in ray casting"""

    def __init__(self, width, height):
        """RayCube constructor"""

        # set dims
        self.width, self.height = width, height

        # create shader
        self.program = glutils.loadShaders(strVS, strFS)

        # cube vertices
        vertices = numpy.array([
                0.0, 0.0, 0.0,
                1.0, 0.0, 0.0,
                1.0, 1.0, 0.0,
                0.0, 1.0, 0.0,
                0.0, 0.0, 1.0,
                1.0, 0.0, 1.0,
                1.0, 1.0, 1.0,
                0.0, 1.0, 1.0
                ], numpy.float32)

        # cube colors
        colors = numpy.array([
                0.0, 0.0, 0.0,
                1.0, 0.0, 0.0,
                1.0, 1.0, 0.0,
                0.0, 1.0, 0.0,
                0.0, 0.0, 1.0,
                1.0, 0.0, 1.0,
                1.0, 1.0, 1.0,
                0.0, 1.0, 1.0
                ], numpy.float32)

        # individual triangles
        indices = numpy.array([
                4, 5, 7,
                7, 5, 6,
                5, 1, 6,
                6, 1, 2,
                1, 0, 2,
                2, 0, 3,
                0, 4, 3,
                3, 4, 7,
                6, 2, 7,
                7, 2, 3,
                4, 0, 5,
                5, 0, 1
```

```python
                 ], numpy.int16)

        self.nIndices = indices.size

        # set up vertex array object (VAO)
        self.vao = glGenVertexArrays(1)
        glBindVertexArray(self.vao)

        #vertex buffer
        self.vertexBuffer = glGenBuffers(1)
        glBindBuffer(GL_ARRAY_BUFFER, self.vertexBuffer)
        glBufferData(GL_ARRAY_BUFFER, 4*len(vertices), vertices, GL_STATIC_DRAW)

        # vertex buffer - cube vertex colors
        self.colorBuffer = glGenBuffers(1)
        glBindBuffer(GL_ARRAY_BUFFER, self.colorBuffer)
        glBufferData(GL_ARRAY_BUFFER, 4*len(colors), colors, GL_STATIC_DRAW);

        # index buffer
        self.indexBuffer = glGenBuffers(1)
        glBindBuffer(GL_ELEMENT_ARRAY_BUFFER, self.indexBuffer);
        glBufferData(GL_ELEMENT_ARRAY_BUFFER, 2*len(indices), indices,
                     GL_STATIC_DRAW)

        # enable attrs using the layout indices in shader
        aPosLoc = 1
        aColorLoc = 2

        # bind buffers:
        glEnableVertexAttribArray(1)
        glEnableVertexAttribArray(2)

        # vertex
        glBindBuffer(GL_ARRAY_BUFFER, self.vertexBuffer)
        glVertexAttribPointer(aPosLoc, 3, GL_FLOAT, GL_FALSE, 0, None)

        # color
        glBindBuffer(GL_ARRAY_BUFFER, self.colorBuffer)
        glVertexAttribPointer(aColorLoc, 3, GL_FLOAT, GL_FALSE, 0, None)
        # index
        glBindBuffer(GL_ELEMENT_ARRAY_BUFFER, self.indexBuffer)

        # unbind VAO
        glBindVertexArray(0)

        # FBO
        self.initFBO()

    def renderBackFace(self, pMatrix, mvMatrix):
        """renders back-face of ray-cube to a texture and returns it"""
        # render to FBO
        glBindFramebuffer(GL_FRAMEBUFFER, self.fboHandle)
        # set active texture
        glActiveTexture(GL_TEXTURE0)
        # bind to FBO texture
        glBindTexture(GL_TEXTURE_2D, self.texHandle)

        # render cube with face culling enabled
        self.renderCube(pMatrix, mvMatrix, self.program, True)
```

```python
            # unbind texture
            glBindTexture(GL_TEXTURE_2D, 0)
            glBindFramebuffer(GL_FRAMEBUFFER, 0)
            glBindRenderbuffer(GL_RENDERBUFFER, 0)

            # return texture ID
            return self.texHandle

    def renderFrontFace(self, pMatrix, mvMatrix, program):
        """render front face of ray-cube"""
        # no face culling
        self.renderCube(pMatrix, mvMatrix, program, False)

    def renderCube(self, pMatrix, mvMatrix, program, cullFace):
        """render cube use face culling if flag set"""

        glClear(GL_COLOR_BUFFER_BIT | GL_DEPTH_BUFFER_BIT)
        # set shader program
        glUseProgram(program)

        # set projection matrix
        glUniformMatrix4fv(glGetUniformLocation(program, b'uPMatrix'),
                           1, GL_FALSE, pMatrix)

        # set modelview matrix
        glUniformMatrix4fv(glGetUniformLocation(program, b'uMVMatrix'),
                           1, GL_FALSE, mvMatrix)

        # enable face culling
        glDisable(GL_CULL_FACE)
        if cullFace:
            glFrontFace(GL_CCW)
            glCullFace(GL_FRONT)
            glEnable(GL_CULL_FACE)

        # bind VAO
        glBindVertexArray(self.vao)

        # animated slice
        glDrawElements(GL_TRIANGLES, self.nIndices, GL_UNSIGNED_SHORT, None)

        # unbind VAO
        glBindVertexArray(0)

        # reset cull face
        if cullFace:
            # disable face culling
            glDisable(GL_CULL_FACE)

    def reshape(self, width, height):
        self.width = width
        self.height = height
        self.aspect = width/float(height)
        # re-create FBO
        self.clearFBO()
        self.initFBO()

    def initFBO(self):
```

```python
        # create frame buffer object
        self.fboHandle = glGenFramebuffers(1)
        # create texture
        self.texHandle = glGenTextures(1)
        # create depth buffer
        self.depthHandle = glGenRenderbuffers(1)

        # bind
        glBindFramebuffer(GL_FRAMEBUFFER, self.fboHandle)

        glActiveTexture(GL_TEXTURE0)
        glBindTexture(GL_TEXTURE_2D, self.texHandle)

        # set parameters to draw the image at different sizes
        glTexParameteri(GL_TEXTURE_2D, GL_TEXTURE_MIN_FILTER, GL_LINEAR)
        glTexParameteri(GL_TEXTURE_2D, GL_TEXTURE_MAG_FILTER, GL_LINEAR)
        glTexParameteri(GL_TEXTURE_2D, GL_TEXTURE_WRAP_S, GL_CLAMP_TO_EDGE)
        glTexParameteri(GL_TEXTURE_2D, GL_TEXTURE_WRAP_T, GL_CLAMP_TO_EDGE)

        # set up texture
        glTexImage2D(GL_TEXTURE_2D, 0, GL_RGBA, self.width, self.height,
                     0, GL_RGBA, GL_UNSIGNED_BYTE, None)

        # bind texture to FBO
        glFramebufferTexture2D(GL_FRAMEBUFFER, GL_COLOR_ATTACHMENT0,
                               GL_TEXTURE_2D, self.texHandle, 0)

        # bind
        glBindRenderbuffer(GL_RENDERBUFFER, self.depthHandle)
        glRenderbufferStorage(GL_RENDERBUFFER, GL_DEPTH_COMPONENT24,
                              self.width, self.height)

        # bind depth buffer to FBO
        glFramebufferRenderbuffer(GL_FRAMEBUFFER, GL_DEPTH_ATTACHMENT,
                                  GL_RENDERBUFFER, self.depthHandle)

        # check status
        status = glCheckFramebufferStatus(GL_FRAMEBUFFER)
        if status == GL_FRAMEBUFFER_COMPLETE:
            pass
            #print "fbo %d complete" % self.fboHandle
        elif status == GL_FRAMEBUFFER_UNSUPPORTED:
            print("fbo %d unsupported" % self.fboHandle)
        else:
            print("fbo %d Error" % self.fboHandle)

        glBindTexture(GL_TEXTURE_2D, 0)
        glBindFramebuffer(GL_FRAMEBUFFER, 0)
        glBindRenderbuffer(GL_RENDERBUFFER, 0)
        return

    def clearFBO(self):
        """clears old FBO"""
        # delete FBO
        if glIsFramebuffer(self.fboHandle):
            glDeleteFramebuffers(int(self.fboHandle))

        # delete texture
```

```
            if glIsTexture(self.texHandle):
                glDeleteTextures(int(self.texHandle))


    def close(self):
        """call this to free up OpenGL resources"""
        glBindTexture(GL_TEXTURE_2D, 0)
        glBindFramebuffer(GL_FRAMEBUFFER, 0)
        glBindRenderbuffer(GL_RENDERBUFFER, 0)
        # delete FBO
        if glIsFramebuffer(self.fboHandle):
            glDeleteFramebuffers(int(self.fboHandle))

        # delete texture
        if glIsTexture(self.texHandle):
            glDeleteTextures(int(self.texHandle))

        # delete render buffer
        """
        if glIsRenderbuffer(self.depthHandle):
            glDeleteRenderbuffers(1, int(self.depthHandle))
        """
        # delete buffers
        """
        glDeleteBuffers(1, self._vertexBuffer)
        glDeleteBuffers(1, &_indexBuffer)
        glDeleteBuffers(1, &_colorBuffer)
        """
```

11.8 体光线投射

接下来，在 RayCastRender 类中实现光线投射算法。算法的核心发生在该类使用的片段着色器内，该类也使用 RayCube 类来帮助生成光线。要查看 raycast.py 的完整代码，请直接跳到 11.9 节。

开始先创建一个 RayCube 对象，并在其构造函数中加载着色器。

```
    def __init__(self, width, height, volume):
        """RayCastRender construction"""

        # create RayCube object
❶       self.raycube = raycube.RayCube(width, height)

        # set dimensions
        self.width = width
        self.height = height
        self.aspect = width/float(height)

        # create shader
❷       self.program = glutils.loadShaders(strVS, strFS)
        # texture
❸       self.texVolume, self.Nx, self.Ny, self.Nz = volume

        # initialize camera
❹       self.camera = Camera()
```

在❶行，构造函数创建了一个 RayCube 对象，用于生成光线。在❷行，加载光线投射使用的着色器。然后在❸行，设置 OpenGL 的三维纹理和尺寸，它们作为一个元组传入 RayCastRender 的构造函数。在❹行，创建一个 Camera 类实例，它将用于设置 OpenGL 透视转换，从而进行三维渲染（这个类与第 10 章中使用的基本相同）。

下面是 RayCastRender 的渲染方法：

```
def draw(self):
    # build projection matrix
❶   pMatrix = glutils.perspective(45.0, self.aspect, 0.1, 100.0)

    # modelview matrix
❷   mvMatrix = glutils.lookAt(self.camera.eye, self.camera.center,
                              self.camera.up)

    # render

    # generate ray-cube back-face texture
❸   texture = self.raycube.renderBackFace(pMatrix, mvMatrix)

    # set shader program
❹   glUseProgram(self.program)

    # set window dimensions
    glUniform2f(glGetUniformLocation(self.program, b"uWinDims"),
                float(self.width), float(self.height))

    # bind to texture unit 0, which represents back-faces of cube
❺   glActiveTexture(GL_TEXTURE0)
    glBindTexture(GL_TEXTURE_2D, texture)
    glUniform1i(glGetUniformLocation(self.program, b"texBackFaces"), 0)

    # texture unit 1: 3D volume texture
❻   glActiveTexture(GL_TEXTURE1)
    glBindTexture(GL_TEXTURE_3D, self.texVolume)
    glUniform1i(glGetUniformLocation(self.program, b"texVolume"), 1)

    # draw front-face of cubes
❼   self.raycube.renderFrontFace(pMatrix, mvMatrix, self.program)
```

在❶行，利用 glutils.perspective()工具方法，建立一个用于渲染的透视投影矩阵。然后，在❷行，将当前摄像机参数设置给 glutils.lookAt()方法。在❸行，完成第一遍渲染，即利用 RayCube 的 renderBackFace()方法，将彩色立方体的背面绘制成一个纹理（该方法也返回生成的纹理的 ID）。

在❹行，为光线投射算法启用着色器。然后在❺行，设置❸行返回的纹理，准备将它用于着色器程序中，作为纹理单元 0。在❻行，设置从读入的体数据中创建的三维纹理，作为纹理单元 1，这样在着色器中就可以用这两个纹理。最后，在❼行，用 RayCube 的 renderFrontFace()方法渲染立方体的正面。执行这段代码后，RayCastRender 着色器将作用于顶点和片段。

11.8.1 顶点着色器

接下来是 RayCastRender 使用的着色器。先看看顶点着色器：

```
#version 330 core
```
❶ ```
layout(location = 1) in vec3 cubePos;
layout(location = 2) in vec3 cubeCol;
```
❷ ```
uniform mat4 uMVMatrix;
uniform mat4 uPMatrix;
```
❸ ```
out vec4 vColor;

void main()
{
    // set position
```
❹ ```
    gl_Position = uPMatrix * uMVMatrix * vec4(cubePos.xyz, 1.0);

    // set color
```
❺ ```
    vColor = vec4(cubeCol.rgb, 1.0);
}
```

从❶行开始，设置位置和颜色的输入变量。布局使用的索引与 RayCube 顶点着色器中定义相同，因为 RayCastRender 使用该类中定义的 VBO 来绘制几何形状，着色器中的位置必须相匹配。在❷行和随后一行中，定义输入变换矩阵。然后，在❸行设置颜色值，作为着色器的输出。❹行是常用变换，计算了内建的 gl_Position 输出。在❺行，将输出设置为立方体顶点的当前颜色，顶点之间将进行插值，从而在片段着色器中给出正确的颜色。

### 11.8.2 片段着色器

片段着色器是重头戏。它实现了光线投射算法的核心。

```
#version 330 core

in vec4 vColor;

uniform sampler2D texBackFaces;
uniform sampler3D texVolume;
uniform vec2 uWinDims;

out vec4 fragColor;

void main()
{
    // start of ray
```
❶ ```
    vec3 start = vColor.rgb;

    // calculate texture coordinates at fragment,
    // which is a fraction of window coordinates
```
❷ ```
    vec2 texc = gl_FragCoord.xy/uWinDims.xy;
```
```

```
        // get end of ray by looking up back-face color
❸       vec3 end = texture(texBackFaces, texc).rgb;

        // calculate ray direction
❹       vec3 dir = end - start;

        // normalized ray direction
        vec3 norm_dir = normalize(dir);

        // the length from front to back is calculated and
        // used to terminate the ray
        float len = length(dir.xyz);

        // ray step size
        float stepSize = 0.01;

        // x-ray projection
        vec4 dst = vec4(0.0);

        // step through the ray
❺       for(float t = 0.0; t < len; t += stepSize) {

            // set position to end point of ray
❻           vec3 samplePos = start + t*norm_dir;

            // get texture value at position
❼           float val = texture(texVolume, samplePos).r;
            vec4 src = vec4(val);

            // set opacity
❽           src.a *= 0.1;
            src.rgb *= src.a;

            // blend with previous value
❾           dst = (1.0 - dst.a)*src + dst;

            // exit loop when alpha exceeds threshold
❿           if(dst.a >= 0.95)
                break;
        }

        // set fragment color
        fragColor = dst;
}
```

片段着色器的输入是立方体顶点的颜色。片段着色器也能获取通过渲染颜色立方体生成的二维纹理，包含数据的三维纹理，以及 OpenGL 窗口的尺寸。

在片段着色器执行时，我们传入立方体的正面，这样通过在 ❶ 行查找传入的颜色值，会得到进入这个立方体的光线的起点（回想一下 11.1.2 小节中关于立方体中的颜色与光线方向的联系的讨论）。

在 ❷ 行，计算屏幕上的传入片段的纹理坐标。用片段在窗口坐标中的位置除以窗口尺寸，得到范围在[0,1]的值。在 ❸ 行，利用该纹理坐标来查找立方体的背面颜色，获得光线的终点。

在❹行，计算光线方向，然后计算光线的法向和长度，它们将用于光线投射计算。然后在❺行，利用光线的起点和方向，直到光线的终点，我们循环通过该物体。在❻行，计算光线在体数据中当前的位置，在❼行，查找这一点上的数据值。

混合公式给出 X 射线效果，在❽和❾行执行。我们结合 dst 值和当前的强度值（它利用 alpha 值衰减），该过程沿光线继续（alpha 值不断增加）。

在❿行，检查这个 alpha 值，直到它等于最大阈值 0.95，就退出循环。最后的结果是某种平均的透明度，穿透物体的每个像素，产生"透视"或 x 射线效果（请尝试改变阈值和 alpha 衰减，产生不同的效果）。

11.9 完整的体光线投射代码

以下是完整的代码清单。也可以在 https:// github.com/electronut/pp/tree/master/volrender/找到 raycast.py 文件。

```
import OpenGL
from OpenGL.GL import *
from OpenGL.GL.shaders import *

import numpy as np
import math, sys

import raycube, glutils, volreader

strVS = """
#version 330 core

layout(location = 1) in vec3 cubePos;
layout(location = 2) in vec3 cubeCol;

uniform mat4 uMVMatrix;
uniform mat4 uPMatrix;

out vec4 vColor;

void main()
{
    // set position
    gl_Position = uPMatrix * uMVMatrix * vec4(cubePos.xyz, 1.0);

    // set color
    vColor = vec4(cubeCol.rgb, 1.0);
}
"""
strFS = """
#version 330 core

in vec4 vColor;

uniform sampler2D texBackFaces;
uniform sampler3D texVolume;
uniform vec2 uWinDims;
```

```
out vec4 fragColor;

void main()
{
    // start of ray
    vec3 start = vColor.rgb;

    // calculate texture coords at fragment,
    // which is a fraction of window coords
    vec2 texc = gl_FragCoord.xy/uWinDims.xy;

    // get end of ray by looking up back-face color
    vec3 end = texture(texBackFaces, texc).rgb;

    // calculate ray direction
    vec3 dir = end - start;

    // normalized ray direction
    vec3 norm_dir = normalize(dir);

    // the length from front to back is calculated and
    // used to terminate the ray
    float len = length(dir.xyz);

    // ray step size
    float stepSize = 0.01;

    // x-ray projection
    vec4 dst = vec4(0.0);

    // step through the ray
    for(float t = 0.0; t < len; t += stepSize) {

        // set position to end point of ray
        vec3 samplePos = start + t*norm_dir;

        // get texture value at position
        float val = texture(texVolume, samplePos).r;
        vec4 src = vec4(val);

        // set opacity
        src.a *= 0.1;
        src.rgb *= src.a;

        // blend with previous value
        dst = (1.0 - dst.a)*src + dst;

        // exit loop when alpha exceeds threshold
        if(dst.a >= 0.95)
            break;
    }

    // set fragment color
    fragColor = dst;
}
"""

class Camera:
    """helper class for viewing"""
    def __init__(self):
```

```python
        self.r = 1.5
        self.theta = 0
        self.center = [0.5, 0.5, 0.5]
        self.eye = [0.5 + self.r, 0.5, 0.5]
        self.up = [0.0, 0.0, 1.0]

    def rotate(self, clockWise):
        """rotate eye by one step"""
        if clockWise:
            self.theta = (self.theta + 5) % 360
        else:
            self.theta = (self.theta - 5) % 360
        # recalculate eye
        self.eye = [0.5 + self.r*math.cos(math.radians(self.theta)),
                    0.5 + self.r*math.sin(math.radians(self.theta)),
                    0.5]

class RayCastRender:
    """class that does Ray Casting"""

    def __init__(self, width, height, volume):
        """RayCastRender constr"""

        # create RayCube object
        self.raycube = raycube.RayCube(width, height)
        # set dimensions
        self.width = width
        self.height = height
        self.aspect = width/float(height)

        # create shader
        self.program = glutils.loadShaders(strVS, strFS)
        # texture
        self.texVolume, self.Nx, self.Ny, self.Nz = volume

        # initialize camera
        self.camera = Camera()

    def draw(self):

        # build projection matrix
        pMatrix = glutils.perspective(45.0, self.aspect, 0.1, 100.0)

        # modelview matrix
        mvMatrix = glutils.lookAt(self.camera.eye, self.camera.center,
                                  self.camera.up)
        # render

        # generate ray-cube back-face texture
        texture = self.raycube.renderBackFace(pMatrix, mvMatrix)

        # set shader program
        glUseProgram(self.program)

        # set window dimensions
        glUniform2f(glGetUniformLocation(self.program, b"uWinDims"),
                    float(self.width), float(self.height))

        # texture unit 0, which represents back-faces of cube
        glActiveTexture(GL_TEXTURE0)
```

```
            glBindTexture(GL_TEXTURE_2D, texture)
            glUniform1i(glGetUniformLocation(self.program, b"texBackFaces"), 0)

            # texture unit 1: 3D volume texture
            glActiveTexture(GL_TEXTURE1)
            glBindTexture(GL_TEXTURE_3D, self.texVolume)
            glUniform1i(glGetUniformLocation(self.program, b"texVolume"), 1)

            # draw front face of cubes
            self.raycube.renderFrontFace(pMatrix, mvMatrix, self.program)

            #self.render(pMatrix, mvMatrix)

    def keyPressed(self, key):
        if key == 'l':
            self.camera.rotate(True)
        elif key == 'r':
            self.camera.rotate(False)

    def reshape(self, width, height):
        self.width = width
        self.height = height
        self.aspect = width/float(height)
        self.raycube.reshape(width, height)

    def close(self):
        self.raycube.close()
```

11.10 二维切片

除了显示体数据的三维视图，我们也想在屏幕上显示 x、y 和 z 方向上的数据的二维切片。这段代码封装在 SliceRender 类中，它创建二维体切片。要查看 slicerender.py 的完整代码，请直接跳到 11.11 节。

下面的初始化代码为切片建立了几何形状：

```
        # set up vertex array object (VAO)
        self.vao = glGenVertexArrays(1)
        glBindVertexArray(self.vao)

        # define quad vertices
❶       vertexData = numpy.array([ 0.0, 1.0, 0.0,
                                    0.0, 0.0, 0.0,
                                    1.0, 1.0, 0.0,
                                    1.0, 0.0, 0.0], numpy.float32)
        # vertex buffer
        self.vertexBuffer = glGenBuffers(1)
        glBindBuffer(GL_ARRAY_BUFFER, self.vertexBuffer)
        glBufferData(GL_ARRAY_BUFFER, 4*len(vertexData), vertexData,
                     GL_STATIC_DRAW)
        # enable arrays
        glEnableVertexAttribArray(self.vertIndex)
        # set buffers
        glBindBuffer(GL_ARRAY_BUFFER, self.vertexBuffer)
```

```
    glVertexAttribPointer(self.vertIndex, 3, GL_FLOAT, GL_FALSE, 0, None)

    # unbind VAO
    glBindVertexArray(0)
```

这段代码设置了一个 VAO 来管理 VBO，像前几个例子一样。在❶行定义的几何形状是在 xy 平面上的正方形（顶点顺序是 GL_TRIANGLE_STRIP 的顺序，第 9 章介绍过），所以无论是否显示垂直于 x、y 或 z 的切片，都使用相同的几何形状。这些情况的不同之处在于，从三维纹理中选择显示的数据平面。讨论顶点着色器时，会回过头来探讨这一点。

接下来，利用 SliceRender 渲染二维切片：

```
def draw(self):
    # clear buffers
    glClear(GL_COLOR_BUFFER_BIT | GL_DEPTH_BUFFER_BIT)
    # build projection matrix
❶   pMatrix = glutils.ortho(-0.6, 0.6, -0.6, 0.6, 0.1, 100.0)
    # modelview matrix
❷   mvMatrix = numpy.array([1.0, 0.0, 0.0, 0.0,
                            0.0, 1.0, 0.0, 0.0,
                            0.0, 0.0, 1.0, 0.0,
                            -0.5, -0.5, -1.0, 1.0], numpy.float32)
    # use shader
    glUseProgram(self.program)

    # set projection matrix
    glUniformMatrix4fv(self.pMatrixUniform, 1, GL_FALSE, pMatrix)

    # set modelview matrix
    glUniformMatrix4fv(self.mvMatrixUniform, 1, GL_FALSE, mvMatrix)

    # set current slice fraction
❸   glUniform1f(glGetUniformLocation(self.program, b"uSliceFrac"),
                float(self.currSliceIndex)/float(self.currSliceMax))
    # set current slice mode
❹   glUniform1i(glGetUniformLocation(self.program, b"uSliceMode"),
                self.mode)

    # enable texture
    glActiveTexture(GL_TEXTURE0)
    glBindTexture(GL_TEXTURE_3D, self.texture)
    glUniform1i(glGetUniformLocation(self.program, b"tex"), 0)

    # bind VAO
    glBindVertexArray(self.vao)
    # draw
    glDrawArrays(GL_TRIANGLE_STRIP, 0, 4)
    # unbind VAO
    glBindVertexArray(0)
```

每个二维切片都是一个正方形，是用 OpenGL 的三角形带图元建立的。这段代码完成了三角形带的渲染设置。请注意，我们用 glutils.ortho() 方法实现了正投影。在❶行，建立了一个投影，在表示切片的单位正方形边上增加 0.1 的缓冲区。用 OpenGL 绘制时，默认视图（不应用任何转换）将眼睛放在（0，0，0），沿着 z 轴向下看而 y 轴朝上。在❷行，对几何形狀应用转换（-0.5，-0.5，-1.0），让它围绕 z 轴居中。在❸行设置当前切片的分数（例如，100 个切片中的第 10 个将是 0.1），在❹行设置切片模式（在 x，y 或 z 方向上观看切片，分别由整数 0、1 和 2 表示），并将这两个值设置给着色器。

11.10.1 顶点着色器

现在来看看 SliceRender 的顶点着色器：

```
# version 330 core

in vec3 aVert;

uniform mat4 uMVMatrix;
uniform mat4 uPMatrix;

uniform float uSliceFrac;
uniform int uSliceMode;

out vec3 texcoord;

void main() {
    // x slice
    if (uSliceMode == 0) {
❶       texcoord = vec3(uSliceFrac, aVert.x, 1.0-aVert.y);
    }
    // y slice
    else if (uSliceMode == 1) {
❷       texcoord = vec3(aVert.x, uSliceFrac, 1.0-aVert.y);
    }
    // z slice
    else {
❸       texcoord = vec3(aVert.x, 1.0-aVert.y, uSliceFrac);
    }

    // calculate transformed vertex
    gl_Position = uPMatrix * uMVMatrix * vec4(aVert, 1.0);
}
```

顶点着色器用三角形带顶点数组作为输入，并设置一个纹理坐标作为输出。当前切片分数和切片模式作为 uniform 变量传入。

在❶行，计算 x 切片的纹理坐标。因为要垂直于 x 方向切片，所以希望切片平行于 yz 平面。传入顶点着色器的三维顶点也兼作三维纹理坐标，因为它们在[0, 1]的范围内，所以纹理坐标由（f, Vx, Vy）给出，其中 f 是 x 轴方向上切片编号的分数，Vx 和 Vy 是顶点坐标。遗憾的是，由此产生的图像是倒的，因为 OpenGL 坐标系统的原点在左下角，y 方向朝上。这和我们希望的相反。为了解决这个问题，将纹理坐标 t 改为（1 - t），采用（f, Vx, 1 - Vy），如❶行所示。在❷行和❸行，利用类似的逻辑来计算 y 和 z 方向上切片的纹理坐标。

11.10.2 片段着色器

下面是片段着色器：

```
# version 330 core
```
❶ `in vec3 texcoord;`

❷ `uniform sampler3D texture;`

```
out vec4 fragColor;

void main() {
    // look up color in texture
```
❸ `    vec4 col = texture(tex, texcoord);`
❹ `    fragColor = col.rrra;`
```
}
```

在❶行，片段着色器声明 texcoord 作为输入，它在顶点着色器中设为输出。在❷行，该纹理采样器声明为 uniform。在❸行，用 texcoord 查找纹理的颜色，在❹行，设置 fragColor 作为输出（因为读入纹理时只作为红色通道，所以使用 col.rrra）。

11.10.3 针对二维切片的用户界面

现在需要为用户提供一种方法，让切片经过这些数据。用 SliceRender 的键盘处理程序来实现。

```
    def keyPressed(self, key):
        """keypress handler"""
        if key == 'x':
❶           self.mode = SliceRender.XSLICE
            # reset slice index
            self.currSliceIndex = int(self.Nx/2)
            self.currSliceMax = self.Nx
        elif key == 'y':
            self.mode = SliceRender.YSLICE
            # reset slice index
            self.currSliceIndex = int(self.Ny/2)
            self.currSliceMax = self.Ny
        elif key == 'z':
            self.mode = SliceRender.ZSLICE
            # reset slice index
            self.currSliceIndex = int(self.Nz/2)
            self.currSliceMax = self.Nz
        elif key == 'l':
❷           self.currSliceIndex = (self.currSliceIndex + 1) % self.currSliceMax
        elif key == 'r':
            self.currSliceIndex = (self.currSliceIndex - 1) % self.currSliceMax
```

在键盘上按下 X、Y 或 Z 键时，SliceRender 切换到的 x、y 或 z 切片模式。在 ❶行，可以看到 x 切片的效果，随后将当前切片索引设置为数据的中间位置，并更新最大切片数。按下键盘上的左、右箭头键，就实现了切片翻页。在 ❷行，按下右箭头时切片索引递增。取模运算符（%）确保在超过最大值时索引"翻转"为 0。

11.11 完整的二维切片代码

以下是完整的代码清单。也可以在 https://github.com/electronut/pp/tree/master/volrender/ 找到 slicerender.py 文件。

```
import OpenGL
from OpenGL.GL import *
from OpenGL.GL.shaders import *
import numpy, math, sys

import volreader, glutils

strVS = """
# version 330 core

in vec3 aVert;

uniform mat4 uMVMatrix;
uniform mat4 uPMatrix;

uniform float uSliceFrac;
uniform int uSliceMode;

out vec3 texcoord;

void main() {

    // x slice
    if (uSliceMode == 0) {
        texcoord = vec3(uSliceFrac, aVert.x, 1.0-aVert.y);
    }
    // y slice
    else if (uSliceMode == 1) {
        texcoord = vec3(aVert.x, uSliceFrac, 1.0-aVert.y);
    }
    // z slice
    else {
        texcoord = vec3(aVert.x, 1.0-aVert.y, uSliceFrac);
    }

    // calculate transformed vertex
    gl_Position = uPMatrix * uMVMatrix * vec4(aVert, 1.0);
}
"""
strFS = """
# version 330 core

in vec3 texcoord;
```

```
uniform sampler3D tex;

out vec4 fragColor;

void main() {
    // look up color in texture
    vec4 col = texture(tex, texcoord);
    fragColor = col.rrra;
}
"""

class SliceRender:
    # slice modes
    XSLICE, YSLICE, ZSLICE = 0, 1, 2

    def __init__(self, width, height, volume):
        """SliceRender constructor"""
        self.width = width
        self.height = height
        self.aspect = width/float(height)

        # slice mode
        self.mode = SliceRender.ZSLICE

        # create shader
        self.program = glutils.loadShaders(strVS, strFS)

        glUseProgram(self.program)

        self.pMatrixUniform = glGetUniformLocation(self.program, b'uPMatrix')
        self.mvMatrixUniform = glGetUniformLocation(self.program,
                                                    b"uMVMatrix")
        # attributes
        self.vertIndex = glGetAttribLocation(self.program, b"aVert")

        # set up vertex array object (VAO)
        self.vao = glGenVertexArrays(1)
        glBindVertexArray(self.vao)

        # define quad vertices
        vertexData = numpy.array([ 0.0, 1.0, 0.0,
                                   0.0, 0.0, 0.0,
                                   1.0, 1.0, 0.0,
                                   1.0, 0.0, 0.0], numpy.float32)
        # vertex buffer
        self.vertexBuffer = glGenBuffers(1)
        glBindBuffer(GL_ARRAY_BUFFER, self.vertexBuffer)
        glBufferData(GL_ARRAY_BUFFER, 4*len(vertexData), vertexData,
                     GL_STATIC_DRAW)
        # enable arrays
        glEnableVertexAttribArray(self.vertIndex)
        # set buffers
        glBindBuffer(GL_ARRAY_BUFFER, self.vertexBuffer)
        glVertexAttribPointer(self.vertIndex, 3, GL_FLOAT, GL_FALSE, 0, None)

        # unbind VAO
        glBindVertexArray(0)
```

```python
        # load texture
        self.texture, self.Nx, self.Ny, self.Nz = volume

        # current slice index
        self.currSliceIndex = int(self.Nz/2);
        self.currSliceMax = self.Nz;


    def reshape(self, width, height):
        self.width = width
        self.height = height
        self.aspect = width/float(height)

    def draw(self):
        # clear buffers
        glClear(GL_COLOR_BUFFER_BIT | GL_DEPTH_BUFFER_BIT)
        # build projection matrix
        pMatrix = glutils.ortho(-0.6, 0.6, -0.6, 0.6, 0.1, 100.0)
        # modelview matrix
        mvMatrix = numpy.array([1.0, 0.0, 0.0, 0.0,
                                0.0, 1.0, 0.0, 0.0,
                                0.0, 0.0, 1.0, 0.0,
                                -0.5, -0.5, -1.0, 1.0], numpy.float32)
        # use shader
        glUseProgram(self.program)

        # set projection matrix
        glUniformMatrix4fv(self.pMatrixUniform, 1, GL_FALSE, pMatrix)

        # set modelview matrix
        glUniformMatrix4fv(self.mvMatrixUniform, 1, GL_FALSE, mvMatrix)

        # set current slice fraction
        glUniform1f(glGetUniformLocation(self.program, b"uSliceFrac"),
                    float(self.currSliceIndex)/float(self.currSliceMax))
        # set current slice mode
        glUniform1i(glGetUniformLocation(self.program, b"uSliceMode"),
                    self.mode)

        # enable texture
        glActiveTexture(GL_TEXTURE0)
        glBindTexture(GL_TEXTURE_3D, self.texture)
        glUniform1i(glGetUniformLocation(self.program, b"tex"), 0)

        # bind VAO
        glBindVertexArray(self.vao)
        # draw
        glDrawArrays(GL_TRIANGLE_STRIP, 0, 4)
        # unbind VAO
        glBindVertexArray(0)

    def keyPressed(self, key):
        """keypress handler"""
        if key == 'x':
            self.mode = SliceRender.XSLICE
            # reset slice index
            self.currSliceIndex = int(self.Nx/2)
            self.currSliceMax = self.Nx
        elif key == 'y':
```

```
                self.mode = SliceRender.YSLICE
                # reset slice index
                self.currSliceIndex = int(self.Ny/2)
                self.currSliceMax = self.Ny
            elif key == 'z':
                self.mode = SliceRender.ZSLICE
                # reset slice index
                self.currSliceIndex = int(self.Nz/2)
                self.currSliceMax = self.Nz
            elif key == 'l':
                self.currSliceIndex = (self.currSliceIndex + 1) % self.currSliceMax
            elif key == 'r':
                self.currSliceIndex = (self.currSliceIndex - 1) % self.currSliceMax

    def close(self):
        pass
```

11.12 代码整合

让我们快速看看项目的主文件 volrender.py。该文件用到 RenderWin 类，它创建并管理 GLFW 的 OpenGL 窗口（这里不会详细介绍该类，因为它类似于第 9 章和第 10 章中使用的类）。要查看 volrender.py 的完整代码，请直接跳到 11.13 节。

在这个类的初始化代码中，创建了渲染器，如下所示：

```
       # load volume data
❶      self.volume = volreader.loadVolume(imageDir)
       # create renderer
❷      self.renderer = RayCastRender(self.width, self.height, self.volume)
```

在❶行，读入三维数据作为 OpenGL 纹理。在❷行，创建 RayCastRender 类型的对象来显示数据。

键盘上按下 V 键，在体渲染和切片渲染之间切换。下面是 RenderWindow 的键盘处理程序：

```
    def onKeyboard(self, win, key, scancode, action, mods):
        # print 'keyboard: ', win, key, scancode, action, mods
        # ESC to quit
        if key is glfw.GLFW_KEY_ESCAPE:
            self.renderer.close()
            self.exitNow = True
        else:
❶           if action is glfw.GLFW_PRESS or action is glfw.GLFW_REPEAT:
                if key == glfw.GLFW_KEY_V:
                    # toggle render mode
❷                   if isinstance(self.renderer, RayCastRender):
                        self.renderer = SliceRender(self.width, self.height,
                                                     self.volume)
                    else:
                        self.renderer = RayCastRender(self.width, self.height,
                                                       self.volume)
                    # call reshape on renderer
```

```
                self.renderer.reshape(self.width, self.height)
            else:
                # send keypress to renderer
❸               keyDict = {glfw.GLFW_KEY_X: 'x', glfw.GLFW_KEY_Y: 'y',
                           glfw.GLFW_KEY_Z: 'z', glfw.GLFW_KEY_LEFT: 'l',
                           glfw.GLFW_KEY_RIGHT: 'r'}
                try:
                    self.renderer.keyPressed(keyDict[key])
                except:
                    pass
```

按 ESC 键退出程序。其他按键（V、X、Y、Z 等）在❶行处理（无论刚刚按下该键，或者保持按下状态，代码都有效）。在❷行，如果按下 V 键，在体渲染和切片渲染之间切换，利用 Python 的 isinstance()方法来确定当前类的类型。

要处理除 ESC 之外的按键事件，我们使用一个字典❸，将按下的键传入渲染器的 keyPressed()处理程序。

注意　　我没有直接传入 glfw.KEY 值，而是利用字典将它们转换为字符值，因为这是很好的做法，目的是减少源文件中的依赖关系。目前，项目中依赖 GLFW 的唯一文件是 volrender.py。如果将 GLFW 相关的类型传入其他代码，它们就需要导入并依赖 GLFW 库，但如果切换到另一个 OpenGL 窗口工具包，代码就乱了。

11.13　完整的主文件代码

以下是完整的代码清单。也可以在 https://github.com/electronut/pp/tree/master/volrender/找到 volrender.py 文件。

```
import sys, argparse, os
from slicerender import *
from raycast import *
import glfw

class RenderWin:
    """GLFW Rendering window class"""
    def __init__(self, imageDir):

        # save current working directory
        cwd = os.getcwd()

        # initialize glfw; this changes cwd
        glfw.glfwInit()

        # restore cwd
        os.chdir(cwd)

        # version hints
```

```python
        glfw.glfwWindowHint(glfw.GLFW_CONTEXT_VERSION_MAJOR, 3)
        glfw.glfwWindowHint(glfw.GLFW_CONTEXT_VERSION_MINOR, 3)
        glfw.glfwWindowHint(glfw.GLFW_OPENGL_FORWARD_COMPAT, GL_TRUE)
        glfw.glfwWindowHint(glfw.GLFW_OPENGL_PROFILE,
                            glfw.GLFW_OPENGL_CORE_PROFILE)

        # make a window
        self.width, self.height = 512, 512
        self.aspect = self.width/float(self.height)
        self.win = glfw.glfwCreateWindow(self.width, self.height, b"volrender")
        # make context current
        glfw.glfwMakeContextCurrent(self.win)

        # initialize GL
        glViewport(0, 0, self.width, self.height)
        glEnable(GL_DEPTH_TEST)
        glClearColor(0.0, 0.0, 0.0, 0.0)

        # set window callbacks
        glfw.glfwSetMouseButtonCallback(self.win, self.onMouseButton)
        glfw.glfwSetKeyCallback(self.win, self.onKeyboard)
        glfw.glfwSetWindowSizeCallback(self.win, self.onSize)

        # load volume data
        self.volume = volreader.loadVolume(imageDir)
        # create renderer
        self.renderer = RayCastRender(self.width, self.height, self.volume)

        # exit flag
        self.exitNow = False

    def onMouseButton(self, win, button, action, mods):
        # print 'mouse button: ', win, button, action, mods
        pass

    def onKeyboard(self, win, key, scancode, action, mods):
        # print 'keyboard: ', win, key, scancode, action, mods
        # ESC to quit
        if key is glfw.GLFW_KEY_ESCAPE:
            self.renderer.close()
            self.exitNow = True
        else:
            if action is glfw.GLFW_PRESS or action is glfw.GLFW_REPEAT:
                if key == glfw.GLFW_KEY_V:
                    # toggle render mode
                    if isinstance(self.renderer, RayCastRender):
                        self.renderer = SliceRender(self.width, self.height,
                                                    self.volume)
                    else:
                        self.renderer = RayCastRender(self.width, self.height,
                                                      self.volume)
                    # call reshape on renderer
                    self.renderer.reshape(self.width, self.height)
                else:
                    # send keypress to renderer
                    keyDict = {glfw.GLFW_KEY_X: 'x', glfw.GLFW_KEY_Y: 'y',
                               glfw.GLFW_KEY_Z: 'z', glfw.GLFW_KEY_LEFT: 'l',
```

```python
                            glfw.GLFW_KEY_RIGHT: 'r'}
            try:
                self.renderer.keyPressed(keyDict[key])
            except:
                pass

    def onSize(self, win, width, height):
        #print 'onsize: ', win, width, height
        self.width = width
        self.height = height
        self.aspect = width/float(height)
        glViewport(0, 0, self.width, self.height)
        self.renderer.reshape(width, height)

    def run(self):
        # start loop
        while not glfw.glfwWindowShouldClose(self.win) and not self.exitNow:
            # render
            self.renderer.draw()
            # swap buffers
            glfw.glfwSwapBuffers(self.win)
            # wait for events
            glfw.glfwWaitEvents()
        # end
        glfw.glfwTerminate()

# main() function
def main():
    print('starting volrender...')
    # create parser
    parser = argparse.ArgumentParser(description="Volume Rendering...")
    # add expected arguments
    parser.add_argument('--dir', dest='imageDir', required=True)
    # parse args
    args = parser.parse_args()

    # create render window
    rwin = RenderWin(args.imageDir)
    rwin.run()

# call main
if __name__ == '__main__':
    main()
```

11.14 运行程序

下面是用斯坦福体数据档案（Stanford Volume Data Archive）[①]的数据来运行应用程序的示例。

```
$ python volrender.py --dir mrbrain-8bit/
```

应该看到如图 11-6 所示的画面。

---

① http://graphics.stanford.edu/data/voldata/

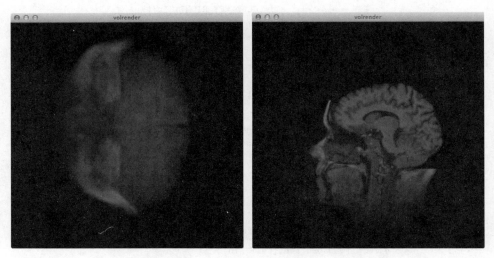

图 11-6　virender.by 运行示例。左侧的图像是体渲染，右侧的图像是二维切片

11.15　小结

本章用 Python 和 OpenGL 实现了体光线投射算法。我们学习了如何用 GLSL 着色器有效地实现这个算法，以及如何从体数据创建二维切片。

11.16　实验

这里有一些方法，让你继续完善体光线投射程序。

1. 目前，在光线投射模式下，很难看到的体数据"立方体"的边界。请实现一个 WireFrame 类，围绕该立方体绘制一个盒子。分别用红、绿、蓝色绘制 x、y 和 z 轴，让它们每个都有自己的着色器。在 RayCastRender 类中使用 WireFrame。

2. 实现数据刻度。在当前的实现中，针对物体绘制了一个正方体，针对二维切片了一个正方形，这假定数据集是对称（即切片的数目在每个方向上相同），但大多数实际数据具有不同数目的切片。特别是医疗数据，常常在 z 方向的切片较少，例如，维度是 256×256×99。要正确显示这些数据，必须在计算中引入刻度。一种做法是将刻度应用于立方体的顶点（三维物体）和正方形的顶点（二维切片）。然后用户可以输入刻度参数作为命令行参数。

3. 我们的体光线投射实现使用了 x 射线投射来计算像素的最终颜色或强度。另一种流行的实现方式是用最大强度投影（maximum intensity projection，MIP），在每个像素上设置最大强度。在代码中实现这一点（提示：在 RayCastRender 的片段着色器中，修改逐步穿透光线的代码，沿着光线检查并设置最大值，代替混合值）。

4. 目前，唯一实现的用户界面是围绕 x 轴、y 轴和 z 轴旋转。请实现缩放功能，按 I 键或 O 键来放大或缩小体渲染图像。可以在 glutils.lookAt()方法中设置适当的相机参数来实现，有一点需要注意：如果将观察位置移动到立方体内部，光线投射将失败，因为 OpenGL 会裁剪立方体的正面。光线投射所需的光线计算，同时需要彩色立方体的正面和背面才能正确渲染。作为替代方案，通过在 glutils.projecton()方法中调节视场来实现缩放。

第五部分

玩转硬件

"系统中的那些可以用锤子击打（不建议这么做）的部分称为硬件，那些只能咒骂的程序指令称为软件。"

——佚名

第 12 章
Arduino 简介

Arduino 是一个简单的微控制器板卡，是基于可编程芯片的开源开发环境。所有版本的 Arduino 都包含计算机的标准组件，如内存、处理器和输入/输出系统。

在本章中，我们将在 Arduino 的帮助下开始进入微控制器的世界。我们将学习 Arduino 平台的基础知识，以及如何用 Arduino 编程语言（C++的一个版本）构建 Arduino 程序。我们将学习如何对 Arduino 进行编程，从一个简单的光传感器电路收集数据，然后通过串行端口将数据发送给计算机。接下来，我们将用 pySerial 通过串口与 Arduino 交互，收集数据，并用 matplotlib 实时绘制图形。新值输入时，图形将向右滚动，就像 EKG 监视器一样。其电路设置如图 12-1 所示。

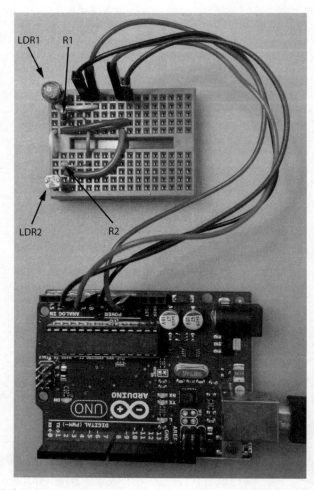

图 12-1　简单的光敏电阻（LDR）电路，在面包板上搭建，并连接到 Arduino Uno

12.1　Arduino

　　Arduino 是围绕一类名为 Atmel AVR 的微控制器芯片构建的平台。市场上有许多不同尺寸和功能的 Arduino 板。图 12-2 突出了 Arduino Uno 板卡的一些主要组件，这是比较常见的、可以获得的板卡。Arduino 板上的接头（如图 12-2 所示）允许你访问微控制器的模拟和数字引脚，以便通过发送和接收数据，与其他电子设备通信。（我推荐阅读 John Boxall 的《Arduino Workshop》一书[No Starch Press，2013]，以便更好地理解 Arduino 的电子和编程知识。）

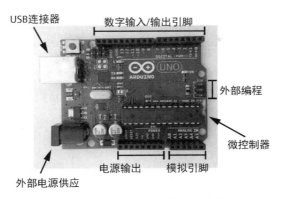

图 12-2　Arduino Uno 板卡的组件

注意　　　虽然本书的项目中使用 Arduino Uno 板，但你应该能够使用非官方的板卡。如果打算这样做，请参阅 Arduino 网站（http://arduino.cc/），了解板卡之间的区别。例如，一些板卡的引脚编号惯例与 Uno 不同。

Arduino Uno 有一个微控制器芯片，一个通用串行总线（USB）连接，一个用于外部电源的插孔，一些数字输入/输出引脚，一些模拟引脚，可供外部电路使用的电源输出，甚至有直接对芯片编程的引脚。

Arduino 板卡包括一个"引导加载程序（bootloader）"，它是一个程序，让你上传代码并在微控制器上运行。如果没有引导加载程序，则要用"在线串行编程技术（ICSP）"与微控制器进行交互（通过名为 ICSP 头的外部编程引脚，Arduino 支持了 ICSP）。Arduino 易于编程，因为可以通过 USB 端口将它连接到计算机，并用 Arduino 软件将代码上传到电路板。

Arduino 板卡最重要的组件是 AVR 微控制器，它是一个单片计算机。Arduino Uno 上的 AVR 微控制器是一个 ATmega328 芯片。它有中央处理单元（CPU）、定时器/计数器、模拟和数字引脚、存储器模块和时钟模块等。芯片的 CPU 执行上传的程序。定时器/计数器模块可用于在程序内创建周期性事件（例如每秒检查某个数字引脚的值）。模拟引脚使用模数转换器（ADC）模块将输入的模拟信号转换为数字值，数字引脚可以作为输入或输出，具体取决于如何设置。

12.2　Arduino 生态系统

Arduino 处在一个生态系统的中心，这个生态系统将编程语言与集成开发环境（IDE）、支持性和创造性的社区以及大量外设结合起来。

12.2.1 语言

Arduino 编程语言是 C++的简化版本,起源于处理和布线(Processing and Wiring)原型语言。它的设计目的是让不熟悉编程的人容易使用。为 Arduino 编写的程序称为"草图(sketches)"(有关 Arduino 编程语言的更多信息,请访问 http://arduino.cc/)。

12.2.2 IDE

Arduino 包括一个简单的 IDE,可以在其中创建草图并上传到 Arduino(参见图 12-3)。IDE 还包括一个"串行监视器(serial monitor)",可以通过让 Arduino 经串口向计算机发送信息,来调试应用程序。此外,IDE 还包括几个示例程序,以及一系列标准库,用于执行常见任务,并与外部的外设板卡接口。

图 12-3 Arduino IDE 中的示例程序

12.2.3 社区

Arduino 有一个庞大的用户群,如果对一个项目有疑问,你可以向 Arduino 社区寻求帮助。Arduino 社区已经开发了许多开源库,可以在你的项目中使用,如果用 Arduino 连接一些传感器模块时遇到困难,很可能有人已经解决了这个问题,并提供了一个库,使你的生活更轻松。

12.2.4 外设

与所有流行的平台一样，围绕 Arduino 平台建立了一个产业。大量的功能扩展板（可以方便地安装在 Arduino 上，以便访问传感器和其他电子元件的板卡）、分线板（可以方便地连接难以焊接的元件/电路）和其他外设，都可用于 Arduino，其中许多可以让你的项目更简单。SparkFun Electronics（https://www.sparkfun.com/）和 Adafruit Industries（http://www.adafruit.com/）是两家公司，它们分别提供了许多可与 Arduino 一起使用的产品。

12.3 所需模块

既然已经知道了 Arduino 的一些基础知识，就让我们看看如何对板卡编程，从光感测电路读取数据。除了 Arduino，我们还需要两个电阻和两个"光敏电阻（LDR）"。电阻用于减少通过电路的电流并降低电压。我们将使用光敏电阻（也称为 photoresistor），其电阻随着光照强度的增加而减小。我们还需要一个面包板和一些电线来搭电路，并用万用表来检查连接。

12.4 搭建感光电路

首先，我们将搭建感光电路，它由两个常规电阻和两个 LDR 组成。图 12-4 展示了感光电路的电路图（附录 B 包含如何开始搭建电子电路的一些基本信息）。

在图 12-4 中，VCC 表示连接到 Arduino 的 5V 输出，它为电路供电。标记为 LDR1 和 LDR2 的元件是两个光敏电阻，A0 和 A1 是 Arduino 的模拟引脚 0 和 1（这些模拟引脚允许微控制器从外部电路读取电压电平）。你还可以看到电阻 R1 和 R2，以及接地（GND）连接，它可以是 Arduino 上的任何 GND 引脚。可以在面包板（带弹簧夹的塑料板，用于在不焊接的情况下组装电路）上搭建电路，用少量电线连接，如图 12-1 所示。

12.4.1 电路工作原理

在该电路中，每个 LDR 和它下面的电阻，构成了"电阻分压器"，这是一个简单电路，用两个电阻将输入电压分为两部分。该设置意味着 A0 处的电压计算如下：

$$V_0 = V \frac{R_1}{R_{LDR} + R_1}$$

R_1 是电阻的阻值，R_{LDR} 是 LDR 的阻值。V 是电源电压，V_0 是 R_1 两端的电压。

随着照在 LDR 上的光强度改变，它的电阻也改变，因此它两端的电压也相应地改变。电压（0 至 5 伏特之间）在 A0 处读取，并作为 10 位值发送给程序，范围为 [0,1023]。即电压范围映射为[0,1023]的整数。

在电路中使用的 R_1 和 R_2 电阻的值取决于选择的 LDR。如果用光照 LDR，其电阻会降低，因此 R_{LDR} 减小，这意味着 V_0 增加，Arduino 从连接到该 LDR 的模拟引脚处读入较高的值。要确定 R_1 和 R_2 所需的电阻值，请在不同光线条件下用万用表测量 LDR 的电阻，并将该值代入电压公式。你希望在使用 LDR 的光照变化场景下，电压有良好的变化（从 0V 到 5V）。

我最后用了 4.7k 欧姆电阻作为 R_1 和 R_2，因为我的 LDR 的电阻从大约 10k 欧姆（在黑暗中）变化到 1k 欧姆（在明亮的光线中）。将这些值代入上一个公式中，会看到模拟输入需要 1.6V 至 4V 的电压范围。构建电路时，可以从 4.7k 欧姆电阻开始，并根据需要更改它们。

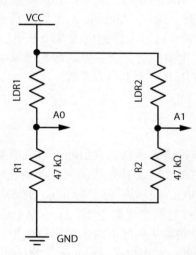

图 12-4　简单感光电路的示意图

12.4.2　Arduino 程序

现在来编写 Arduino 程序。下面是运行在 Arduino 上的代码，让它从电路读取信号，并通过串口将它们发送给计算机：

```
#include "Arduino.h"

void setup()
{
    // initialize serial communications
❶   Serial.begin(9600);
}
```

```
    void loop()
    {
        // read A0
❷      int val1 = analogRead(0);
        // read A1
❸      int val2 = analogRead(1);
        // print to serial
❹      Serial.print(val1);
        Serial.print(" ");
        Serial.print(val2);
        Serial.print("\n");
        // wait
❺      delay(50);
    }
```

在❶行，在 setup()方法中启用串口通信。因为 setup()仅在程序启动时调用，所以这是放置所有初始化代码的好地方。这里将串口通信的波特率（速度，以位/秒为单位）初始化为 9600，这是大多数设备的默认值，对你的任务来说也足够快了。

主要代码位于 loop()方法中。在❷行，从模拟引脚 0（范围在[0,1023]中的一个 10 位整数）读取当前信号值，在❸行，从引脚 1 读取当前信号值。在❹行和随后几行，用 Serial.print()将值发送给计算机，格式化为两个整数，以空格分隔，后跟换行符。在❺行，用 delay()延缓操作指定的时间，在这里是 50 毫秒，然后循环重复。这里采用的值决定了 AVR 微控制器执行 loop()方法的速度。

为了将程序上传到 Arduino，将 Arduino 连接到计算机，启动其 IDE，并启动一个新项目。然后在程序窗口中输入代码，单击 Verify 以编译代码。IDE 将显示所有和语法相关的错误与警告。如果一切正常，单击 Upload 将草图发送给 Arduino。如果在这个阶段没有看到任何错误，从 Arduino 软件的工具菜单中调出串口监视器，应该看到类似这样的结果：

```
512 300
513 280
400 200
...
```

这些是从模拟引脚 0 和 1 读取的模拟值，并通过 Arduino 的 USB 口串行发送给计算机。

12.4.3 创建实时图表

为了在项目中实现滚动的实时图，需要使用一个 deque，如第 4 章所述。deque 由 N 个值的数组构成，在任一端添加和删除值都能很快完成。新值出现时，它们被添加到 deque，并且最老的值被弹出。以固定间隔绘制这些值，就会生成实时图，其中最新的数据总是在左侧添加。

12.5 Python 代码

现在来看看从串口读取数据的 Python 程序（完整的项目代码，请跳到 12.6 节）。为了更好地组织代码，我们定义一个类 AnalogPlot，它保存要绘制的数据。下面是该类的构造函数：

```
class AnalogPlot:
    # constructor
    def __init__(self, strPort, maxLen):
        # open serial port
❶       self.ser = serial.Serial(strPort, 9600)

❷       self.a0Vals = deque([0.0]*maxLen)
❸       self.a1Vals = deque([0.0]*maxLen)
❹       self.maxLen = maxLen
```

在❶行，AnalogPlot 的构造函数利用 pySerial 库创建一个 Serial 对象。我们将用该类与 Arduino 的串口通信。Serial 构造函数的第一个参数是端口名称字符串，可以在 IDE 中通过选择 Tools▶Serial Port 找到它（在 Windows 上，该字符串类似于 COM3，在 Linux 和 OS X 上，类似于/dev/tty.usbmodem411）。Serial 构造函数的第二个参数是波特率，我们设置为 9600，以匹配 Arduino 程序中设置的速率。

在❷行和❸行，创建了 deque 对象来保存模拟值。我们使用大小为 maxLen 的全零列表初始化 deque 对象，这是在给定时间绘制的值的最大数量。在❹行，这个 maxLen 保存在 AnalogPlot 对象中。

为了实时绘制模拟值，我们用 deque 对象缓冲最近的值，如 AnalogPlot 类中所示。

```
    # add data
    def add(self, data):
        assert(len(data) == 2)
❶       self.addToDeq(self.a0Vals, data[0])
❷       self.addToDeq(self.a1Vals, data[1])

    # add to deque; pop oldest value
    def addToDeq(self, buf, val):
❸       buf.pop()
❹       buf.appendleft(val)
```

正如你前面看到的，Arduino 每行仅发送两个模拟整数值。在 add()方法中，在❶行和❷行，每个模拟引脚的数据值都用 addToDeq()方法添加到两个 deque 对象中。在❸行和❹行，这个方法用 pop()方法从 deque 的尾部删除最老的值，然后用 appendleft()方法将最新的值添加到 deque 的头部。在绘制来自该 deque 的值时，最新的值始终显示在图形的左侧。

我们利用 matplotlib 的 animation 类，以设置的间隔更新绘图（你可能还记得，在第 5 章的 Boids 项目中看到过）。下面是 AnalogPlot 中的 update()方法，它将在绘制动画的每一步中调用：

```
# update plot
def update(self, frameNum, a0, a1):
    try:
❶       line = self.ser.readline()
❷       data = [float(val) for val in line.split()]
        # print data
        if(len(data) == 2):
❸           self.add(data)
❹           a0.set_data(range(self.maxLen), self.a0Vals)
❺           a1.set_data(range(self.maxLen), self.a1Vals)
    except:
❻       pass

    return a0, a1
```

在❶行，update()方法将一行串口数据读入为一个字符串，在❷行，利用 Python 的列表解析，将值转换为浮点数，并保存在列表中。我们用 split()方法根据空格来分割字符串，以便将串口读入的字符串 512 600\n 转换为[512,600]。

检查数据有两个值之后，在❸行用 AnalogPlot 的 add()方法将值添加到 deque。在❹行和❺行，利用 matplotlib 的 set_data()方法，用新值更新图形。每次绘图的 x 值是一系列数字[0，... maxLen]，用 range()方法设置。y 值用更新的 deque 对象来填充。

所有这些代码都包含在 try 语句块中，如果发生异常，代码跳转到❻行的 pass，在这里忽略读入数据（pass 不做任何事情）。使用 try 语句块是因为，串口数据有时可能因电路中接触不良而损坏，而我们不希望仅仅因为串口发送了一些坏的值，程序就崩溃。

准备退出时，关闭串口以释放所有系统资源，如下所示：

```
# clean up
def close(self):
    # close serial
    self.ser.flush()
    self.ser.close()
```

在 main()方法中，需要设置 matplotlib 动画：

```
    # set up animation
❶   fig = plt.figure()
❷   ax = plt.axes(xlim=(0, maxLen), ylim=(0, 1023))
❸   a0, = ax.plot([], [])
❹   a1, = ax.plot([], [])
❺   anim = animation.FuncAnimation(fig, analogPlot.update,
                        fargs=(a0, a1), interval=20)

    # show plot
❻   plt.show()
```

在❶行，取得 matplotlib 的 figure 模块，其中包含所有的绘图元素。在❷行，访问 axes 模块，设置图形的 x 和 y 限制。x 限制是样本数，y 限制为 1023，因为它是模拟值范围的上限。

在❸行和❹行，创建两个空白行对象（a0 和 a1），我们将它们传递给 animation 类来设置回调，为每一行提供坐标。然后，在❺行，传入在每个动画步骤中要调用的 analogPlot update()方法，来设置动画。我们还指定了调用该方法的参数，以及以毫秒为单位的时间间隔，这里是 20。在❻行，调用 plt.show()来启动动画。

程序的 main()方法也是用 Python 模块 argparse 来支持命令行选项的地方。

```
# create parser
parser = argparse.ArgumentParser(description="LDR serial")
# add expected arguments
parser.add_argument('--port', dest='port', required=True)
parser.add_argument('--N', dest='maxLen', required=False)

# parse args
args = parser.parse_args()

strPort = args.port

# plot parameters
maxLen = 100
if args.maxLen:
    maxLen = int(args.maxLen)
```

--port 参数是必需的。它告诉程序接收数据的串口名称（在 Arduino IDE 中的 Tools ▸ Serial Port 下找到）。maxLen 参数是可选的，用于设置一次绘制的点数（默认值为 100 个样本）。

12.6 完整的 Python 代码

下面是这个项目的完整 Python 代码。也可以从 https://github.com/electronut/pp/tree/master/arduino-ldr/ldr.py 下载完整的代码。

```
import serial, argparse
from collections import deque
import matplotlib.pyplot as plt
import matplotlib.animation as animation

# plot class
class AnalogPlot:
    # constructor
    def __init__(self, strPort, maxLen):
        # open serial port
        self.ser = serial.Serial(strPort, 9600)

        self.a0Vals = deque([0.0]*maxLen)
        self.a1Vals = deque([0.0]*maxLen)
        self.maxLen = maxLen
```

```python
        # add data
        def add(self, data):
            assert(len(data) == 2)
            self.addToDeq(self.a0Vals, data[0])
            self.addToDeq(self.a1Vals, data[1])

        # add to deque; pop oldest value
        def addToDeq(self, buf, val):
            buf.pop()
            buf.appendleft(val)

        # update plot
        def update(self, frameNum, a0, a1):
            try:
                line = self.ser.readline()
                data = [float(val) for val in line.split()]
                # print data
                if(len(data) == 2):
                    self.add(data)
                    a0.set_data(range(self.maxLen), self.a0Vals)
                    a1.set_data(range(self.maxLen), self.a1Vals)
            except:
                pass

            return a0, a1

        # clean up
        def close(self):
            # close serial
            self.ser.flush()
            self.ser.close()

# main() function
def main():
    # create parser
    parser = argparse.ArgumentParser(description="LDR serial")
    # add expected arguments
    parser.add_argument('--port', dest='port', required=True)
    parser.add_argument('--N', dest='maxLen', required=False)
    # parse args
    args = parser.parse_args()

    #strPort = '/dev/tty.usbserial-A7006Yqh'
    strPort = args.port

    print('reading from serial port %s...' % strPort)

    # plot parameters
    maxLen = 100
    if args.maxLen:
        maxLen = int(args.maxLen)

    # create plot object
    analogPlot = AnalogPlot(strPort, maxLen)

    print('plotting data...')

    # set up animation
    fig = plt.figure()
    ax = plt.axes(xlim=(0, maxLen), ylim=(0, 1023))
```

```
a0, = ax.plot([], [])
a1, = ax.plot([], [])
anim = animation.FuncAnimation(fig, analogPlot.update,
                               fargs=(a0, a1), interval=20)

# show plot
plt.show()

# clean up
analogPlot.close()

print('exiting.')

# call main
if __name__ == '__main__':
    main()
```

12.7 运行程序

要测试该程序，请组装 LDR 电路，将 Arduino 连接到计算机，上传程序，然后运行 Python 代码。

```
$ python3 --port /dev/tty.usbmodem411 ldr.py
```

图 12-5 展示了程序的输出示例，具体来说，是 LDR 暴露在光下然后覆盖时生成的图。从图中可以看出，当 LDR 的电阻变化时，Arduino 读取的模拟电压也发生变化。中心的高峰发生在我迅速将手放在 LDR 上时，右侧平坦部分发生在我较慢地移过手时。

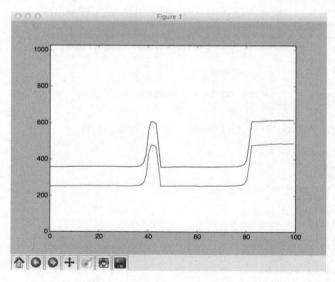

图 12-5　光敏绘图程序的运行示例

这两个 LDR 具有不同的电阻特性，这就是两条线不完全重合的原因，但是你可以看到，它们以相同的方式对光强的变化做出反应。

12.8 小结

这个项目介绍了微控制器和 Arduino 平台的世界。我们学习了 Arduino 编程语法以及如何将程序上传到 Arduino，还学习了如何从 Arduino 引脚读取模拟值，并创建一个简单的 LDR 电路。此外，我们学习了如何通过串口从 Arduino 发送数据，并使用 Python 从计算机读取数据，还学习了使用实时滚动图和 matplotlib 让数据可视化。

12.9 实验

请尝试对 Arduino 项目进行如下修改。

1. 在程序中，图形从左到右滚动，换言之，新值从左边进来时，较老的值向右移动。请反转滚动方向，使图形从右向左移动。

2. Arduino 代码定期读取模拟值，并将它们发送到串口。输入数据可能在某些类型的传感器中存在波动，常见的做法是应用某种类型的滤波，从而平滑数据。请实现 LDR 数据的均值策略（提示：保存每个 LDR 读取的 N 个模拟值的移动平均值，平均值应定期发送到串口。减少循环中的 delay()，以便更快地读取值。均值图比原始图更平滑吗？尝试不同的 N 值，看看会发生什么）。

3. 传感器电路中有两个 LDR。保持这些 LDR 在良好的光源下，从它们上面用手扫过。你应该看到图形的急剧变化：一个 LDR 曲线在另一个之前变化，因为它首先被遮挡。可以用这些信息来检测手的运动方向吗？这是一个基本的手势检测项目（提示：LDR 中图形的急剧变化发生在不同的时间，说明哪个 LDR 先被遮蔽，从而给出手的运动方向）。

第 13 章
激光音乐秀

在第 12 章中，我们学习了 Arduino 的基础知识，这很适合与底层电子设备接口。在本项目中，我们将利用 Arduino 搭建硬件，通过音频信号产生有趣的激光图案。这一次，Python 会承担更重的任务。除了处理串口通信之外，它还会基于实时音频数据进行计算，并利用该数据来调整激光显示设备中的电机。

出于这些目的，请将激光当作一束强烈的光束，即使通过很大距离投射，它仍保持聚焦在一个微小的点上。这种聚焦是可能的，因为光束被组织成光波只沿一个方向行进，且彼此同相。这个项目将用一个便宜的、容易获得的激光笔来创建激光图案，与音乐同步（或任何音频输入）。我们用激光笔和两个连接到电机的旋转镜，来构建生成有趣图案的硬件。我们用 Arduino 设置电机的方向和旋转速度，利用 Python 通过串口来控制。Python 程序将读取音频输入，分析它，并将它转换为电机速度和方向数据来控制电机。我们还将学习如何设置电机的速度和方向，让图案与音乐同步。

本项目将进一步提升你的 Arduino 和 Python 知识。以下是将要介绍的一些主题：

- 用激光和两个旋转镜产生有趣的图案；

- 用快速傅立叶变换从信号中获得频率信息；
- 用 numpy 计算快速傅里叶变换；
- 用 pyaudio 读取音频数据；
- 在计算机和 Arduino 之间设置串口通信；
- 用 Arduino 驱动电机。

13.1 用激光产生图案

为了在此项目中生成激光图案，我们用激光笔和两个镜子连接到两个小型直流电机的轴上，如图 13-1 所示。如果你平面镜（反射镜 A）的表面照射激光，即使电机正在旋转，投影的反射将保持为一个点。因为激光器的反射平面垂直于电机的旋转轴，所以就像镜子根本不旋转一样。

现在，假设镜子与轴成一定角度连接，如图 13-1 右侧所示（反射镜 B）。当轴旋转时，投影点的轨迹是一个椭圆，而且如果电机旋转足够快，观察者会感觉到移动点是连续的形状。

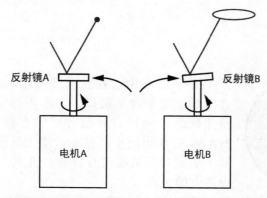

图 13-1 平面镜（镜子 A）反射单个点。倾斜的镜子（镜子 B）的反射在电动机旋转时产生一个圆

如果安排镜子，使从镜子 A 反射的光投影到镜子 B 上，会怎么样？现在当电机 A 和电机 B 旋转时，由反射点产生的图案将是电机 A 和电机 B 的两个旋转运动的组合，产生出有趣的图案，如图 13-2 所示。

产生的图案将取决于两个电机的旋转速度和旋转方向，但是它们类似于在第 2 章中探讨的万花尺产生的长短辐圆内旋轮线。

13.1.1 电机控制

我们用 Arduino 来控制电机的速度和方向。这种设置要小心，确保它可以接受电机相对较高的电压，因为 Arduino 只能承担这么大的电流，否则会损坏。可以用

图 13-3（a）所示的 SparkFun TB6612FNG 外设"分线（breakout）"板来保护 Arduino，简化设计并缩短开发时间。利用分线板从 Arduino 同时控制两个电机。

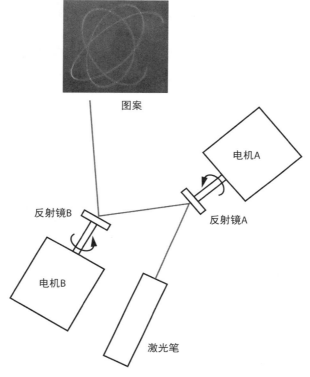

图 13-2　反射激光离开两面旋转的、倾斜的镜子，产生了有趣的复杂图案

(a)　　　　　　　　　(b)

图 13-3　SparkFun 电机驱动器 1A 双通道 TB6612FNG

图 13-3（b）展示了分线板的焊接背面。引脚名称中的 A 和 B 表示两个电机。

第 13 章　激光音乐秀　231

IN 引脚控制电机的方向，01 和 02 引脚为电机供电，PWM 引脚控制电机速度。通过写入这些引脚，你可以控制每个电机的旋转方向和速度，而这正是本项目需要的。

> **注意** 可以用任何你熟悉的电机控制电路替换该分线板部分，只要适当地修改 Arduino 程序即可。

13.1.2 快速傅里叶变换

因为本项目的最终目标是基于音频输入控制电机速度，所以需要能够分析音频的频率。

回顾一下第 4 章，来自乐器的音调是多种频率或泛音的混合。事实上，任何声音都可以用傅立叶变换分解为成分频率。将傅里叶变换应用于数字信号时，结果称为离散傅里叶变换（DFT），因为数字信号由许多离散样本组成。在本项目中，我们用 Python 来实现快速傅立叶变换（FFT）算法，计算 DFT（在本章中，我将使用 FFT 来指代算法和结果）。

下面是一个简单的 FFT 示例。图 13-4 展示了一个只包含两个正弦波的信号，下面是对应的 FFT。上面的波可以用以下等式表示，是两个波相加：

$$y(t) = 4\sin(2\pi 10t) + 2.5\sin(2\pi 30t)$$

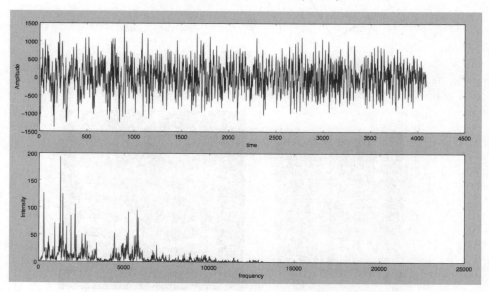

图 13-4　从音乐中记录的音频信号（上）及相应的 FFT（下）

请注意第一个波的表达式中的 4 和 10：4 是波的振幅，10 是波的频率（以赫兹为单位）。同时，第二个波的振幅是 2.5，频率是 30。

FFT 揭示了波的分量频率及其相对振幅，显示峰值在 10 Hz 和 30 Hz。第一个峰值的强度约为第二个峰值的强度的两倍。

现在来看一个更复杂的例子。图 13-5 展示了上面的音频信号，以及下面相应的 FFT。

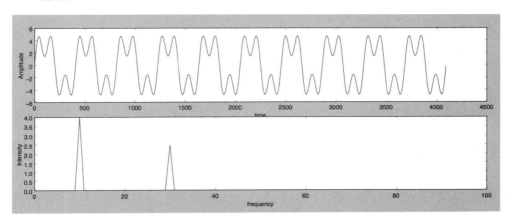

图 13-5　FFT 算法利用振幅信号（上）计算其分量频率（下）

音频输入（或信号）处在"时域"中，因为振幅数据随时间变化。FFT 处在"频域"中。注意图 13-5 中，FFT 显示了一系列峰值，显示了信号中各种频率的强度。

要计算 FFT，需要一组样本。样本数的选择是有点随意的，但是小的样本规模不能给出信号频率特征的良好图景，还可能意味着更大的计算量，因为每秒需要计算更多的 FFT。另一方面，过大的样本规模让信号的变化平均化，因此不会得到信号的"实时"频率响应。对于本项目采用的 44100Hz 的采样率，2048 的样本大小表示大约 0.046 秒的数据。

对于本项目，我们需要将音频数据分解成它的组成频率，并用该信息控制电机。首先，将频率范围（以 Hz 为单位）分成 3 个频段：[0,100]、[100,1000]和[1000,2500]。我们将计算每个频带的平均振幅，每个值将不同程度地影响电机和产生的激光图案，如下所示：

- 低频平均振幅的变化将影响第一个电机的速度；
- 中频平均振幅的变化将影响第二个电机的速度；
- 高频峰值高于某个阈值时，第一个电机将改变方向。

13.2　所需模块

以下是构建此项目所需物品的列表：

- 一支小激光笔；

- 两个直流电动机,如用于小玩具(额定为 9V)的电机;
- 两个小镜子,直径约 2.54 厘米或更小;
- SparkFun 电机驱动器 1A 双通道 TB6612FNG;
- Arduino Uno 或类似板卡;
- 连接的电线(单芯连接线,两侧的公插针工作良好);
- 四节 AA 电池组;
- 一些乐高模块,将电机和激光笔从安装板上抬起,使镜子可以自由旋转;
- 一个矩形的纸板或丙烯酸树脂板,约 20.3 厘米×15.2 厘米,以安装硬件;
- 一把热胶枪;
- 烙铁。

13.2.1 搭建激光秀

第一件事是将镜子连接到电机。反射镜必须与电机轴成一个小角度。要安装镜子,将其面朝下放置在平坦的表面上,并在中心滴一滴热胶。小心地将电机轴浸入胶水中,使其垂直于镜子,直到胶水硬化(如图 13-6 所示)。要测试它,就用手旋转镜子,同时用激光笔照射它。你应该发现反射的激光点投射在平坦表面上时,移动轨迹是一个椭圆中。对第二面镜子也一样操作。

图 13-6　以微小角度将反射镜连接到每个电机轴

对准镜子

接下来,将激光笔与反射镜对准,使激光从反射镜 A 反射到 B,如图 13-7 所示。确保在反射镜 A 的整个旋转范围内,来自反射镜 A 的反射激光保持在反射镜 B 的圆周内(这需要一些尝试,可能会犯些错)。要测试,请手动旋转反射镜 A。另外,确保在两个反射镜的旋转范围内,反射镜 B 的位置让它反射的光落在一个平面上(如墙上)。

图 13-7　激光笔和反射镜的对准

注意　调整时,需要保持激光笔打开。如果激光笔有一个开启按钮,请用胶带让它保持打开(或者参见 13.8 节,了解更优雅的控制激光笔电源的方法)。

对镜子的位置感到满意后,将激光笔和两个带有镜子的电机用热熔胶粘到 3 个一样的乐高模块上,将它们升起,让它们能够自由旋转。接下来,将模块放在安装板上,当你对它们的摆放感到满意时,通过用铅笔记录它们的边缘,标记每个模块的位置。然后将模块粘在板上。

为电机供电

如果电机没有附带连接到它们的端子上的电线(大多数没有),就将电线焊接到两个端子,确保留出足够的电线(例如 15 厘米),以便将电机连接到电机驱动板卡上。电机由电池组中的四节 AA 电池供电,可以用热熔胶将它粘在安装板的背面,如图 13-8 所示。

现在用手旋转两个镜子来测试硬件,同时用激光照射它们。如果足够快地旋转它们,应该看到一些有趣的图案,窥见结果的样子!

图 13-8　将电池组固定在安装板背面

13.2.2　连接电机驱动器

本项目用 Arduino 通过 Sparkfun 电机驱动器（TB6612FNG）来控制电机。我不会详细介绍这个板卡的工作原理，但如果你好奇，可以从了解"H 桥"开始，这是一种常见的电路设计，用"金属氧化物半导体场效应晶体管（MOSFET）"来控制电机。

现在将电机连接到 SparkFun 电机驱动器和 Arduino。有不少电线要连接，如表 13-1 所示。一个电机标记为 A，另一个为 B，在接线时遵守这个惯例。

表 13-1　SparkFun 电机驱动器到 Arduino 的接线

从	至
Arduino Digital Pin 12 TB6612FNG	Pin BIN2
Arduino Digital Pin 11 TB6612FNG	Pin BIN1
Arduino Digital Pin 10 TB6612FNG	Pin STBY
Arduino Digital Pin 9 TB6612FNG	Pin AIN1
Arduino Digital Pin 8 TB6612FNG	Pin AIN2
Arduino Digital Pin 5 TB6612FNG	Pin PWMB
Arduino Digital Pin 3 TB6612FNG	Pin PWMA
Arduino 5V Pin TB6612FNG	Pin VCC
Arduino GND TB6612FNG	Pin GND
Arduino GND Battery Pack	GND（-）

续表

从	至
Battery Pack VCC (+) TB6612FNG	Pin VM
Motor #1 Connector #1 (polarity doesn't matter) TB6612FNG	Pin A01
Motor #1 Connector #2 (polarity doesn't matter) TB6612FNG	Pin A02
Motor #2 Connector #1 (polarity doesn't matter) TB6612FNG	Pin B01
Motor #2 Connector #2 (polarity doesn't matter) TB6612FNG	Pin B02
Arduino USB connector	计算机的 USB 口

图 13-9 展示了所有接线。

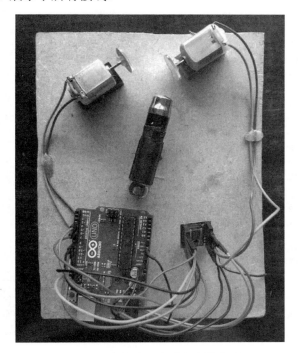

图 13-9 完全连好线的激光秀装置

现在来看看 Arduino 程序。

13.3　Arduino 程序

程序开始时，设置 Arduino 的数字输出引脚。然后，在主循环中，从串口读入数据，并将数据转换为需要发送到电机驱动器板卡的参数。我们还要了解如何实现电机的速度和方向控制。

13.3.1 配置 Arduino 数字输出引脚

首先，根据表 13-1 将 Arduino 数字引脚映射到电机驱动器上的引脚，并将引脚设置为输出。

```
// motor A connected to A01 and A02
// motor B connected to B01 and B02
❶ int STBY = 10; //standby

// Motor A
int PWMA = 3; //speed control
int AIN1 = 9; //direction
int AIN2 = 8; //direction

// Motor B
int PWMB = 5; //speed control
int BIN1 = 11; //direction
❷ int BIN2 = 12; //direction

void setup(){
❸   pinMode(STBY, OUTPUT);

    pinMode(PWMA, OUTPUT);
    pinMode(AIN1, OUTPUT);
    pinMode(AIN2, OUTPUT);

    pinMode(PWMB, OUTPUT);
    pinMode(BIN1, OUTPUT);
    pinMode(BIN2, OUTPUT);

    // initialize serial communication
❹   Serial.begin(9600);
}
```

从❶行到❷行，将 Arduino 引脚的名称映射到电机驱动器引脚。例如，PWMA（脉冲调制 A）控制电机 A 的速度，并分配给 Arduino 引脚 3。PWM 是为设备供电的一种方式，通过发送快速打开和关闭的数字脉冲，让器件"看到"连续的电压。数字脉冲接通的时间部分称为"占空比"，以百分比表示。通过更改这个百分比，可以为设备提供不同的功率水平。PWM 通常用于控制可调光 LED 和电机速度。

然后，在❸行调用 setup()方法，在后续几行中，将所有 7 个数字引脚设置为输出。在❹行，开始串口通信，读取由 Arduino 上的计算机发送的串行数据。

13.3.2 主循环

程序中的主循环等待串行数据到达，解析它以提取电机速度和方向，并利用该信息来设置控制电机的驱动板的数字输出。

```
// main loop that reads the motor data sent by laser.py
void loop()
{
    // data sent is of the form 'H' (header), speed1, dir1, speed2, dir2
❶   if (Serial.available() >= 5) {
❷       if(Serial.read() == 'H') {
            // read the next 4 bytes
❸           byte s1 = Serial.read();
            byte d1 = Serial.read();
            byte s2 = Serial.read();
            byte d2 = Serial.read();

            // stop the motor if both speeds are 0
❹           if(s1 == 0 && s2 == 0) {
                stop();
            }
            else {
                // set the motors' speed and direction
❺               move(0, s1, d1);
                move(1, s2, d2);
            }
            // slight pause for 20 ms
❻           delay(20);
        }
        else {
            // if there is invalid data, stop the motors
❼           stop();
        }
    }
    else {
        // if there is no data, pause for 250 ms
❽       delay(250);
    }
}
```

电机控制数据以 5 个字节为一组发送：H 后跟 4 个单字节数字 s1、d1、s2 和 d2，表示电机的速度和方向。由于串行数据连续输入，所以在❶行，检查并确保已收到至少 5 个字节。如果没有，就延迟 250 毫秒❽，并尝试在下一个周期中再次读取数据。

在❷行，检查读入的第一个字节是一个 H，以确保在一组正确的控制数据的开始处，接下来的 4 个字节是期望的数据。如果不是，就在❼行停止电机，因为数据可能因传输或连接错误而被损坏。

从❸行开始，程序读取两个电机的速度和方向数据。如果两个电机速度都设置为零，就停止电机❹。如果不是，则在❺行用 move() 方法将速度和方向值分配给电机。在❻行，在数据读取中添加一个小延迟，让电机能跟上，确保没有太快地读入数据。

下面是用于设置电机速度和方向的 move() 方法：

```
// set motor speed and direction
// motor: A -> 1, B -> 0
// direction: 1/0
void move(int motor, int speed, int direction)
{
```

```
        // disable standby
❶       digitalWrite(STBY, HIGH);

❷       boolean inPin1 = LOW;
        boolean inPin2 = HIGH;

❸       if(direction == 1){
            inPin1 = HIGH;
            inPin2 = LOW;
        }

        if(motor == 1){
❹           digitalWrite(AIN1, inPin1);
            digitalWrite(AIN2, inPin2);
            analogWrite(PWMA, speed);
        }
        else{
❺           digitalWrite(BIN1, inPin1);
            digitalWrite(BIN2, inPin2);
            analogWrite(PWMB, speed);
        }
    }
```

电机驱动器具有待机模式，以便在电机关闭时节省电力。在❶行，通过写入 HIGH 到待机引脚，退出待机。在❷行，定义两个布尔变量，它们确定电机的旋转方向。在❸行，如果 direction 参数设置为 1，则翻转这些变量的值，这让你在下面的代码中切换电机的方向。

在❹行，为电机 A 设置引脚 AIN1、AIN2 和 PWMA。引脚 AIN1 和 AIN2 控制电机的方向，根据需要，我们用 Arduino 的 digitalWrite()方法将一个引脚设置为 HIGH（1），一个设置为 LOW（0）。对于引脚 PWMA，我们发送 PWM 信号，如前所述，这可以控制电机的速度。要控制 PWM 的值，可以用 analogWrite()方法将范围[0,255]内的值写入 Arduino 输出引脚（不同的是，digitalWrite()方法只允许将 1 或 0 写入输出引脚）。

在❺行，设置电机 B 的引脚。

13.3.3 停止电机

要停止电机，将 LOW 写入电机驱动器的待机引脚。

```
void stop(){
    //enable standby
    digitalWrite(STBY, LOW);
}
```

13.4 Python 代码

现在来看看在计算机上运行的 Python 代码。这段代码完成了繁重的工作：它读入音频，计算 FFT，并发送串行数据到 Arduino。可以在 13.5 节中找到完整的项目代码。

13.4.1 选择音频设备

首先，需要利用 pyaudio 模块读入音频数据。初始化 pyaudio 模块如下：

```
p = pyaudio.PyAudio()
```

接下来，可以用 pyaudio 中的辅助函数访问计算机的音频输入设备。getInputDevice() 方法的代码如下所示：

```
# get pyaudio input device
def getInputDevice(p):
❶    index = None
❷    nDevices = p.get_device_count()
     print('Found %d devices. Select input device:' % nDevices)
     # print all devices found
     for i in range(nDevices):
❸        deviceInfo = p.get_device_info_by_index(i)
❹        devName = deviceInfo['name']
❺        print("%d: %s" % (i, devName))
     # get user selection
     try:
         # convert to integer
❻        index = int(input())
     except:
         pass

     # print the name of the chosen device
     if index is not None:
         devName = p.get_device_info_by_index(index)["name"]
         print("Input device chosen: %s" % devName)
❼    return index
```

在❶行，将 index 变量设置为 None（该 index 是❼行的函数返回值，如果返回为 None，则表明找不到合适的输入设备）。在❷行，用 get_device_count() 方法获取计算机上音频设备的数量，包括所有音频硬件，如麦克风、线路输入或线路输出。然后遍历所有找到的设备，获取每个设备的信息。

❸行的 get_device_info_by_index() 函数返回一个字典，其中包含每个音频设备的各种特征信息，但我们只想查看设备的名称，因为我们正在查找输入设备。在❹行，保存设备名称，在❺行，打印出设备的索引和名称。在❻行，利用 input() 方法读入用户的选择，将读入的字符串转换为整数索引。在❼行，这个选定的索引从函数返回。

13.4.2 从输入设备读取数据

选择输入设备后，需要从中读取数据。为此，我们先打开音频流，如下所示（注意，所有的代码在 while 循环中连续执行）。

```
            # set FFT sample length
❶           fftLen = 2**11
            # set sample rate
❷           sampleRate = 44100

            print('opening stream...')
❸           stream = p.open(format = pyaudio.paInt16,
                            channels = 1,
                            rate = sampleRate,
                            input = True,
                            frames_per_buffer = fftLen,
                            input_device_index = inputIndex)
```

在❶行，将 FFT 缓冲区的长度（用于计算 FFT 的音频采样数）设置为 2048（是 211，FFT 算法针对 2 的幂进行了优化）。然后，将 pyaudio 的采样率设置为 44100，即 44.1 kHz ❷，这是 CD 质量录音的标准。

接下来，打开 pyaudio 流❸，并指定几个选项：

- pyaudio.paInt16 表示读取的数据作为 16 位整数；
- channels 设置为 1，因为我们将音频作为单个频道读取；
- rate 设置为选定的采样速率 44100 Hz；
- input 设置为 True；
- frames_per_buffer 设置为 FFT 缓冲区长度；
- input_device_index 设置为我们在 getInputDevice()方法中选择的设备。

13.4.3　计算数据流的 FFT

以下是从流中读取数据的代码：

```
            # read a chunk of data
❶           data = stream.read(fftLen)
            # convert the data to a numpy array
❷           dataArray = numpy.frombuffer(data, dtype=numpy.int16)
```

在❶行，从音频输入流读取最近的 fftLen 样本。然后在❷行，将该数据转换为 16 位整数 numpy 数组。

现在计算这个数据的 FFT。

```
            # get FFT of data
❶           fftVals = numpy.fft.rfft(dataArray)*2.0/fftLen
            # get absolute values of complex numbers
❷           fftVals = numpy.abs(fftVals)
```

在❶行，用 numpy fft 模块中的 rfft()方法，计算 numpy 数组中的值的 FFT。该方法接受由 "实数"（如音频数据）组成的信号并计算 FFT，通常结果是一组 "复数"。2.0/fftLen 是归一化因子，用来将 FFT 值映射到希望的范围。然后，因为 rfft()方法返回复数，所以使用 numpy 的 abs()方法❷来获取这些复数的大小，它是实数。

13.4.4 从 FFT 值提取频率信息

接下来，从 FFT 值中提取相关的频率信息。

```
# average 3 frequency bands: 0-100 Hz, 100-1000 Hz, and 1000-2500 Hz
levels = [numpy.sum(fftVals[0:100])/100,
          numpy.sum(fftVals[100:1000])/900,
          numpy.sum(fftVals[1000:2500])/1500]
```

为了分析音频信号，我们将频率范围分为 3 个频段：0 至 100 Hz、100 至 1000 Hz 和 1000 至 2500 Hz。我们最感兴趣的是较低的低音频段（0-100 Hz）和中音（100-1000 Hz）频段，分别大致对应于一首歌曲中的节拍和人声。对于每个范围，我们在代码中用 numpy.sum() 方法计算平均 FFT 值。

13.4.5 将频率转换为电机速度和方向

现在将该频率信息转换为电机速度和方向。

```
                # 'H' (header), speed1, dir1, speed2, dir2
❶               vals = [ord('H'), 100, 1, 100, 1]
                # speed1
❷               vals[1] = int(5*levels[0]) % 255
                # speed2
❸               vals[3] = int(100 + levels[1]) % 255

                # dir
                d1 = 0
❹               if levels[2] > 0.1:
                    d1 = 1
                vals[2] = d1
❺               vals[4] = 0
```

在❶行，初始化要发送到 Arduino 的电机速度和方向值的列表（5 个字节，从 H 开始，前面讨论过）。用内置 ord() 函数将字符串转换为整数，然后将 3 个频带的平均值转换为电机速度和方向，填充该列表。

注意　　这部分确实是自由发挥：没有特别优雅的规则来管理这些转换。这些值随着音频信号而不断变化，你提出的任何方法都可以改变电机速度，并随音乐一起影响激光模式。只需确保转换将电机速度设置在[0,255]范围内，并且方向总是设置为 1 或 0。我选择的方法只是基于试验和错误，我在播放各种类型的音乐时观察 FFT 值。

在❷行，从最低频率范围获取值，放大 5 倍，转换为整数，并用取模运算符（%）确保该值位于[0,255]范围内。该值控制第一电机的速度。在❸行，将中间频率值加上 100，并放在[0,255]范围内。该值控制第二电机的速度。

然后，在❹行，只要最高频率范围的值超过阈值 0.1，就切换电机 A 方向。电机 B 的方向保持为 0❺（通过尝试和错误，我发现，这些方法产生很好的图案变化，但我建议你改变这些值，并创建自己的转换。这里没有错误的答案）。

13.4.6　测试电机设置

在用实时音频流测试硬件之前，先检查电机设置。这里展示的 autoTest()函数就是做这件事的：

```
# automatic test for sending motor speeds
def autoTest(ser):
    print('starting automatic test...')
    try:
        while True:
            # for each direction combination
❶          for dr in [(0, 0), (1, 0), (0, 1), (1, 1)]:
                # for a range of speeds
❷              for j in range(25, 180, 10):
❸                  for i in range(25, 180, 10):
❹                      vals = [ord('H'), i, dr[0], j, dr[1]]
❺                      print(vals[1:])
❻                      data = struct.pack('BBBBB', *vals)
❼                      ser.write(data)
                       sleep(0.1)
    except KeyboardInterrupt:
        print('exiting...')
        # shut off motors
❽      vals = [ord('H'), 0, 1, 0, 1]
        data = struct.pack('BBBBB', *vals)
        ser.write(data)
        ser.close()
```

该方法通过改变每个电机的速度和方向，让两个电机在一定范围内运动。因为每个电动机的方向可以是顺时针或逆时针，所以在❶行的外层循环中表示 4 个这样的组合。对于每种组合，❷行和❸行的循环以不同的速度运转电机。

注意　我使用范围（25，180，10），这意味着速度从 25 变到 180，步长为 10。我没有用电机的完整运动范围[0，255]，因为电机很少转速低于 25，而 200 以上的转速真的很快。

在❹行，生成 5 字节的电机数据值，在❺行，打印它们的方向和速度值（使用 Python 字符串切割 vals[1:]将获得除列表中的第一个元素之外的所有内容）。

在❻行，将电机数据打包到字节数组中，并在❼行将其写入串口。按 Ctrl-C 键中断这个测试，在❽行通过清理来处理这个异常，停止电机并关闭串口，像你这样负责的程序员都会这么做。

13.4.7 命令行选项

与以前的项目一样,可以用 argparse 模块来解析程序的命令行参数。

```
# main method
def main():
    # parse arguments
    parser = argparse.ArgumentParser(description='Analyzes audio input and sends motor control information via serial port')
    # add arguments
    parser.add_argument('--port', dest='serial_port_name', required=True)
    parser.add_argument('--mtest', action='store_true', default=False)
    parser.add_argument('--atest', action='store_true', default=False)
    args = parser.parse_args()
```

在这段代码中,串口是必需的命令行选项。还有两个可选的命令行选项:一个用于自动测试(前面介绍过),另一个用于手动测试(稍后将讨论)。

下面是 main() 方法中解析命令行选项后发生的情况:

```
    # open serial port
    strPort = args.serial_port_name
    print('opening ', strPort)
❶   ser = serial.Serial(strPort, 9600)
    if args.mtest:
        manualTest(ser)
    elif args.atest:
        autoTest(ser)
    else:
❷       fftLive(ser)
```

在❶行,利用 pySerial 用传入程序的字符串打开一个串口。串行通信的速度或波特率设置为每秒 9600 位。如果没有使用其他命令参数(--atest 或--mtest),则在❷行继续音频处理和 FFT 计算,这封装在 fftLive() 方法中。

13.4.8 手动测试

这个手动测试允许你输入特定的电机方向和速度,以便看到它们对激光图案的影响。

```
# manual test of motor direction and speeds
def manualTest(ser):
    print('starting manual test...')
    try:
        while True:
            print('enter motor control info such as < 100 1 120 0 >')
❶           strIn = raw_input()
❷           vals = [int(val) for val in strIn.split()[:4]]
❸           vals.insert(0, ord('H'))
❹           data = struct.pack('BBBBB', *vals)
❺           ser.write(data)
    except:
        print('exiting...')
        # shut off the motors
```

```
❻          vals = [ord('H'), 0, 1, 0, 1]
            data = struct.pack('BBBBB', *vals)
            ser.write(data)
            ser.close()
```

在❶行，用 raw_input()方法等待，直到用户在命令提示符下输入值。预期输入的形式为 100 1 120 0，表示电机 A 的速度和方向，然后是电机 B 的速度和方向。在❷行，将字符串解析为整数列表。在❸行，插入一个"H"构成完整的电机数据，在❹行和❺行，打包此数据并通过串口以预期的格式发送。如果用户用 Ctrl-C 中断测试（或者发生任何异常），就在❻行完成清理工作，正常关闭电机和串口。

13.5 完整的 Python 代码

下面是本项目的完整的 Python 代码。也可以在 https://github.com/electronut/pp/tree/master/arduino-laser/laser.py 找到它。

```python
import sys, serial, struct
import pyaudio
import numpy
import math
from time import sleep
import argparse

# manual test of motor direction speeds
def manualTest(ser):
    print('staring manual test...')
    try:
        while True:
            print('enter motor control info: eg. < 100 1 120 0 >')
            strIn = raw_input()
            vals = [int(val) for val in strIn.split()[:4]]
            vals.insert(0, ord('H'))
            data = struct.pack('BBBBB', *vals)
            ser.write(data)
    except:
        print('exiting...')
        # shut off motors
        vals = [ord('H'), 0, 1, 0, 1]
        data = struct.pack('BBBBB', *vals)
        ser.write(data)
        ser.close()

# automatic test for sending motor speeds
def autoTest(ser):
    print('staring automatic test...')
    try:
        while True:
            # for each direction combination
            for dr in [(0, 0), (1, 0), (0, 1), (1, 1)]:
                # for a range of speeds
                for j in range(25, 180, 10):
                    for i in range(25, 180, 10):
```

```python
                            vals = [ord('H'), i, dr[0], j, dr[1]]
                            print(vals[1:])
                            data = struct.pack('BBBBB', *vals)
                            ser.write(data)
                            sleep(0.1)
    except KeyboardInterrupt:
        print('exiting...')
        # shut off motors
        vals = [ord('H'), 0, 1, 0, 1]
        data = struct.pack('BBBBB', *vals)
        ser.write(data)
        ser.close()

# get pyaudio input device
def getInputDevice(p):
    index = None
    nDevices = p.get_device_count()
    print('Found %d devices. Select input device:' % nDevices)
    # print all devices found
    for i in range(nDevices):
        deviceInfo = p.get_device_info_by_index(i)
        devName = deviceInfo['name']
        print("%d: %s" % (i, devName))
    # get user selection
    try:
        # convert to integer
        index = int(input())
    except:
        pass

    # print the name of the chosen device
    if index is not None:
        devName = p.get_device_info_by_index(index)["name"]
        print("Input device chosen: %s" % devName)
    return index

# FFT of live audio
def fftLive(ser):
    # initialize pyaudio
    p = pyaudio.PyAudio()

    # get pyAudio input device index
    inputIndex = getInputDevice(p)

    # set FFT sample length
    fftLen = 2**11
    # set sample rate
    sampleRate = 44100

    print('opening stream...')
    stream = p.open(format = pyaudio.paInt16,
                    channels = 1,
                    rate = sampleRate,
                    input = True,
                    frames_per_buffer = fftLen,
                    input_device_index = inputIndex)
```

```python
    try:
        while True:
            # read a chunk of data
            data = stream.read(fftLen)
            # convert to numpy array
            dataArray = numpy.frombuffer(data, dtype=numpy.int16)
            # get FFT of data
            fftVals = numpy.fft.rfft(dataArray)*2.0/fftLen
            # get absolute values of complex numbers
            fftVals = numpy.abs(fftVals)
            # average 3 frequency bands: 0-100 Hz, 100-1000 Hz and 1000-2500 Hz
            levels = [numpy.sum(fftVals[0:100])/100,
                      numpy.sum(fftVals[100:1000])/900,
                      numpy.sum(fftVals[1000:2500])/1500]

            # the data sent is of the form:
            # 'H' (header), speed1, dir1, speed2, dir2
            vals = [ord('H'), 100, 1, 100, 1]

            # speed1
            vals[1] = int(5*levels[0]) % 255
            # speed2
            vals[3] = int(100 + levels[1]) % 255

            # dir
            d1 = 0
            if levels[2] > 0.1:
                d1 = 1
            vals[2] = d1
            vals[4] = 0

            # pack data
            data = struct.pack('BBBBB', *vals)
            # write data to serial port
            ser.write(data)
            # a slight pause
            sleep(0.001)
    except KeyboardInterrupt:
        print('stopping...')
    finally:
        print('cleaning up')
        stream.close()
        p.terminate()
        # shut off motors
        vals = [ord('H'), 0, 1, 0, 1]
        data = struct.pack('BBBBB', *vals)
        ser.write(data)
        # close serial
        ser.flush()
        ser.close()

# main method
def main():
    # parse arguments
    parser = argparse.ArgumentParser(description='Analyzes audio input and sends motor control information via serial port')
    # add arguments
```

```
    parser.add_argument('--port', dest='serial_port_name', required=True)
    parser.add_argument('--mtest', action='store_true', default=False)
    parser.add_argument('--atest', action='store_true', default=False)
    args = parser.parse_args()

    # open serial port
    strPort = args.serial_port_name
    print('opening ', strPort)
    ser = serial.Serial(strPort, 9600)
    if args.mtest:
        manualTest(ser)
    elif args.atest:
        autoTest(ser)
    else:
        fftLive(ser)

# call main function
if __name__ == '__main__':
    main()
```

13.6 运行程序

为了测试本项目，请组装硬件，将 Arduino 连接到计算机，并将电机驱动程序代码上传到 Arduino。确保电池组已连接，激光笔已打开并投射在墙壁这样的平坦表面上。我建议首先通过运行以下程序来测试激光显示部分（不要忘记更改串口字符串以符合你的计算机）！

```
$ python3 laser.py --port /dev/tty.usbmodem411 --atest
('opening ', '/dev/tty.usbmodem1411')
staring automatic test...
[25, 0, 25, 0]
[35, 0, 25, 0]
[45, 0, 25, 0]
...
```

该测试通过速度和方向的各种组合来运转两个电机。你应该看到投射到墙上的不同激光图案。要停止程序和电机，请按 Ctrl-C。

如果测试成功，就可以开始真正的激光秀了。在计算机上开始播放你喜欢的音乐，然后运行程序如下。（同样，注意串口字符串！）

```
$ python3 laser.py --port /dev/tty.usbmodem411
('opening ', '/dev/tty.usbmodem1411')
Found 4 devices. Select input device:
0: Built-in Microph
1: Built-in Output
2: BoomDevice
3: AirParrot
0
Input device chosen: Built-in Microph
opening stream...
```

你应该看到激光秀产生了许多有趣的图案，随着音乐和时间变化，如图 13-10 所示。

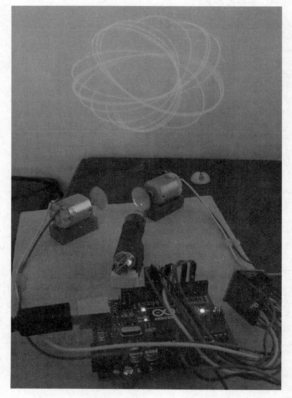

图 13-10　激光秀的完整布线和投影在墙上的图案

13.7　小结

本章通过构建一个较复杂的项目，来提升你的 Python 和 Arduino 技能。你学习了如何使用 Python 和 Arduino 控制电机，用 numpy 来获得音频数据的 FFT，控制串口通信，甚至是激光！

13.8　实验

下面是修改此项目的一些方法。

1. 该程序采用一种自由发挥的方案将 FFT 值转换为电机速度和方向数据。请尝试修改此方案。例如，尝试使用不同的频带和标准来改变电机的方向。

2. 在本项目中，从音频信号收集的频率信息被转换为电机速度和方向。请尝试根据音乐的整体"节奏"或音量来让电机运动。为此，你可以计算信号幅度的"均方根（RMS）"值。该计算类似于 FFT 计算。读入一组音频数据并放入 numpy 数组 x 后，可以按如下方式计算 RMS 值：

```
rms = numpy.sqrt(numpy.mean(x**2))
```

此外，回忆一下，项目中振幅表示为 16 位有符号整数，其最大值为 32768（一个要记住的有用数字，用于标准化）。用这个 RMS 振幅与 FFT 结合，让激光图案产生更大的变化。

在项目中，我们比较朴素地用了一些胶带来保持激光笔的打开状态，测试并进行硬件设置。能找到更好的方法来控制激光吗？请阅读光隔离器和继电器的相关材料[①]，这是可以打开和关闭外部电路的设备。要使用这些设备，首先需要改造激光笔，让它可以通过外部开关切换。一种方法是永久地将激光笔的按钮粘接到 ON 位置，取出电池，并将两个引线焊接到电池触点上。现在就可以用这些电线和激光笔的电池手动打开和关闭激光笔。接下来，通过继电器或光隔离器连接激光笔，并使用 Arduino 上的数字引脚将其打开，用数字开关替换此方案。如果使用光隔离器，可以直接用 Arduino 切换激光开关。如果使用继电器，还需要一个驱动器，通常是一个简单的晶体管电路。

有这样的设置后，请添加一些代码，让 Python 程序运行时，发送一个串行命令到 Arduino，在开始之前打开激光笔。

---

① "Relays and Optoisolators," What-When-How, http://what-when-how.com/8051-microcontroller/relays-and-optoisolators/。

第14章
基于树莓派的天气监控器

当你发现需要更多的计算能力或外设支持的时候，如 USB 或高清晰度多媒体接口（HDMI）视频，你已经离开了像 Arduino 这样的微控制器领域，进入了计算机领域。树莓派（Raspberry Pi）是一个小型计算机，可以很好地执行这样的高级任务，尤其是与 Arduino 相比。

像 Arduino 一样，树莓派在许多有趣的项目中使用。虽然可以放在手掌中，但它是一个完整的计算机（可以连接一个显示器和键盘），这让它在教师和创造者中很受欢迎。

本章将用树莓派以及温度和湿度传感器（DHT11），来构建基于 Web 的温度和湿度监控系统。在树莓派上运行的代码将启动 Bottle Web 服务器，监听传入的连接。当你访问本地网络上的树莓派的 Internet 协议（IP）地址时，Bottle 服务器会提供包含天气数据图表的网页。树莓上的处理程序将与 DHT11 传感器通信，取得数据，并返回给客户程序，客户程序用 flot 库在浏览器中绘制传感器数据。我们还会为连接到树莓派的发光二极管（LED）提供基于 Web 的控制（这只是为了演示如何用树莓派通过 Web 来控制外部设备）。

> **注意** 这个项目使用 Python 2.7。树莓派上的 Raspbian 操作系统提供了 Python 2.7（在 shell 上运行 python）和 Python 3（在 shell 上运行 python3）。编写的代码与两者兼容。

14.1 硬件

像现在销售的所有笔记本电脑或台式机一样，树莓派具有中央处理单元（CPU）、随机存取存储器（RAM）、USB 端口、视频输出、音频输出和网络连接。但与大多数计算机不同，树莓派很便宜：约 35 美元。此外，由于板载的通用输入/输出（GPIO）引脚，树莓派可以轻松与外部硬件接口，让它成为各种嵌入式硬件项目的理想选择。我们将用 DHT11 温度和湿度传感器连接它，监控环境。

14.1.1 DHT11 温湿度传感器

DHT11（见图 14-1）是一种测量温度和湿度的常用传感器。它有 4 个引脚：VDD（+）、GND（-）、DATA，第 4 个不使用。DATA 引脚连接到微控制器（在本例中为树莓派），输入和输出都通过该引脚。我们将用 Adafruit Python 库 Adafruit_Python_DHT 与 DHT11 通信，取得温度和湿度数据。

图 14-1　DHT11 温湿度传感器

14.1.2 树莓派

写作本书时，有 3 种型号的树莓派：Raspberry Pi 1 Model A +、Raspberry Pi 1 Model B +和 Raspberry Pi 2 Model B。这个项目使用了较老的 Raspberry Pi Model B Rev 2，但项目中使用的引脚号码在所有型号中都兼容，并且代码能在所有型号上工作，无需更改。

图 14-2 展示了一个树莓派 B 型计算机，它有两个 USB 端口，一个 HDMI 连接器，一个复合视频输出插孔，一个音频输出插孔，一个用于供电的微型 USB 端口，一个以太网端口和 26 个 GPIO 引脚，引脚排成两列，每列 13 针。板的底侧还有 SD 卡插槽（未显示）。树莓派使用 Broadcom BCM2835 芯片，该芯片有一个 ARM CPU，运行频率为 700 MHz，功耗非常低（这就是为什么树莓派不像台式计算机那样，需要巨大的散热片来冷却它）。B 型还有 512MB 的 SDRAM。这些细节大多数对于这个项目并不重要，但如果你想用最新袖珍计算机的所有规格给人留下深刻印象，这些细节可能很方便。

图 14-2　树莓派型号 B

14.1.3 设置树莓派

与 Arduino 不同，不能将树莓派插入计算机并开始编码。作为一个完备的计算机，树莓派需要一个操作系统和一些外设。至少，建议使用如图 14-3 所示的外设。

- 一张 8GB 或更高容量的 SD 卡，具有合适的操作系统；
- 与树莓派兼容的 USB Wi-Fi 适配器；
- 与树莓派兼容的电源（官方建议是使用 5V 1200 mA 电源，如果需要使用所有 USB 端口，则使用 5V 2500 mA 电源）；
- 保护宝贵的树莓派的盒子；
- 键盘和鼠标（方便起见，请考虑仅占用一个 USB 端口的无线组合）；
- 复合视频电缆或 HDMI 电缆（关于树莓派使用 HDMI 电缆的详细信息，请参阅附录 C）。

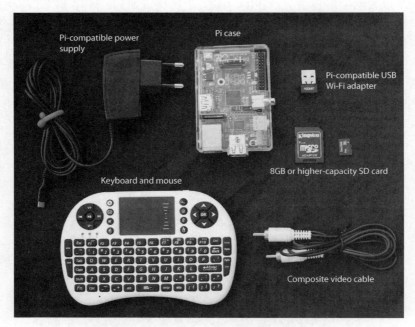

图 14-3　推荐的树莓派外设套件

注意　　购买前，请务必检查已知与 Pi 兼容的外围设备列表，网址在 http://elinuxorg/RPI_VerifiedPeripherals。

14.2　安装和配置软件

现在可以设置树莓派并准备编写 Python 了。下一节将简要介绍所需的步骤，但你应该在安装前，看看 Raspberry Pi Foundation 的"Getting Started with NOOBS"页面上提供的设置视频（http://www.raspberrypi.org/help/noobs-setup/）。

14.2.1 操作系统

树莓派的操作系统和文件将驻留在外部 SD 卡上。虽然有几个操作系统可以选择，但我建议安装 Raspbian。操作系统需要以特定的方式安装，我不会在这里介绍细节（细节可能会改变），而是请你去看在 Linux wiki 上的"RPi Easy SD Card Setup"，网址是 http://elinux.org/RPi_Easy_SD_Card_Setup，或者同一链接中的"Using NOOBS"，它更适合初学者。

14.2.2 初始配置

安装操作系统后，就可以首次启动了。插入格式化的 SD 卡，将复合视频电缆连接到电视或显示器，连接键盘和鼠标，然后将树莓派连接到电源。树莓派启动时，一个叫 raspi-config 的程序应该启动，你应该看到各种配置选项（可以在 Raspberry Pi Foundation 网站上找到 raspi-config 的文档 https://www.raspberrypi.org/documentation/configuration/raspi-config.md）。修改配置如下：

1. 选择 Expand Filesystem 以使用完整的 SD 卡；
2. 选择 Enable Boot to Desktop/Scratch，然后选择 Desktop；
3. 选择 Change Time Zone，并在 Internationalization Options 下设置时区；
4. 转到 Advanced Options 并启用 Overscan；
5. 转到 SSH，通过选择 Enable or Disable SSH Server 选项，在 Advanced Options 菜单中启用远程命令行访问。

现在选择 Finish，树莓派应该重新启动并显示桌面。

14.2.3 Wi-Fi 设置

我们将在此项目中用无线网络连接树莓派。假设安装了兼容的 Wi-Fi 适配器（请先查阅 http://elinux.org/ 网站），Raspbian 应该在插入时自动识别适配器。假设一切正常，我们将设置一个静态 IP 地址，即利用内置的 Nano 编辑器来编辑网络配置文件（Nano 有一个极简单的 UI，熟悉它可能需要一点时间。最重要的事情是记得按 Ctrl-X 并输入 Yes，保存文件并退出）。

我们将在终端中运行命令。打开 LXTerminal（它应该与 Raspbian 一同安装），并输入以下内容：

```
$ sudo nano /etc/network/interfaces
```

该命令将打开 interfaces 文件，我们用它来配置网络设置，如下所示：

```
auto lo

iface lo inet loopback
iface eth0 inet dhcp
```

```
allow-hotplug wlan0
iface wlan0 inet manual
wpa-roam /etc/wpa_supplicant/wpa_supplicant.conf
iface default inet static
address 192.x.x.x
netmask 255.255.255.0
gateway 192.x.x.x
```

添加或修改文件末尾的 address、netmask 和 gateway 等行，以适合本地网络。输入网络的网络掩码（可能为 255.255.255.0），输入网络的网关（在 Linux 上从 Linux 终端运行 ifconfig。在 Windows 上按 Windows 和 R 键，然后运行 ipconfig/all。或者在 OS X 上选择 System Preferences->Network），给树莓派一个静态网络 IP 地址，它不同于网络上任何其他设备的 IP 地址。

现在使用 WiFi Config 实用程序，让树莓派连接 Wi-Fi 网络。你应该在桌面上看到一个快捷方式（如果遇到困难，请尝试 https://learn.adafruit.com/ 上的 Adafruit 教程。如果一切顺利，树莓派应该能用内置的浏览器 Midori 连上因特网）。

14.2.4 设置编程环境

接下来，我们安装开发环境，包括与外部硬件通信所需的 RPi.GPIO 包，Bottle Web 框架以及在 Raspberry Pi 上安装其他 Python 包所需的工具。确保已连接到因特网，并在终端中运行以下命令，每次一条：

```
$ sudo apt-get update
$ sudo apt-get install python-setuptools
$ sudo apt-get install python-dev
$ sudo apt-get install python-rpi.gpio
$ sudo easy_install bottle
```

现在，从 http://www.flotcharts.org/ 下载最新版本的 flot JavaScript 绘图库，将它展开以创建 flot 目录，并将此目录复制到你的程序所在的文件夹中。

```
$ wget http://www.flotcharts.org/downloads/flot-x.zip
$ unzip flot-x.zip
$ mv flot myProjectDir/
```

接下来，在终端中运行以下命令，安装 Adafruit_Python_DHT 库（https://github.com/adafruit/Adafruit_Python_DHT/），你将用它从连接到树莓派的 DHT11 传感器取得数据：

```
$ git clone https://github.com/adafruit/Adafruit_Python_DHT.git
$ cd Adafruit_Python_DHT
$ sudo python setup.py install
```

树莓派现在应该已经设置好，我们已经有了构建天气监视程序所需的全部软件。

14.2.5 通过 SSH 连接

与连接显示器、用鼠标和键盘控制它相比，通过桌面或笔记本电脑登录树莓派是更容易的方法。Linux 和 OS X 内置了这种支持，即 Secure Shell（SSH）。如果你用的是 Windows，请安装 PuTTY 以连接到树莓派。

以下列表展示了典型的 SSH 会话：

```
❶ moksha:~ mahesh$ ssh pi@192.168.4.32
❷ pi@192.168.4.32's password:
❸ pi@raspberrypi ~ $ whoami
  pi
  pi@raspberrypi ~ $
```

在这个会话中，在❶行，输入 ssh 命令、默认用户名（pi）以及 IP 地址，形如 ssh username@ip_address，从我的计算机登录到树莓派。输入 ssh 时❷，会提示输入密码。默认密码为 raspberry。

你认为登录树莓派成功后，请输入 whoami 命令❸。如果响应是 pi，如前所示，则已经正确登录。

注意　更改树莓派的用户名和密码，让它更安全，这是一个好主意。有关使用树莓派远程操作的更多提示，请参阅附录 C。

14.2.6 Web 框架 Bottle

要通过 Web 界面监视和控制树莓派，需要让它运行一个 Web 服务器。我们将使用 Bottle，它是一个具有简单界面的 Python Web 框架（实际上，整个库由名为 bottle.py 的单个源文件组成）。以下是用 Bottle 提供一个简单网页所需的代码：

```
from bottle import route, run

@route('/hello')
def hello():
    return "Hello Bottle World!"

run(host='192.168.x.x', port=xxxx, debug=True)
```

这段代码用 Python 装饰器@route 定义了一条路由，代表一个 URL 或路径，客户端用它来发送数据请求。定义的路由调用路由函数，返回一个字符串。run()方法启动 Bottle 服务器，现在可以接受来自客户端的连接（请务必提供你自己的 IP 地址和端口号）。请注意，我已将 debug 标记设置为 True，这样更容易诊断问题）。

在连接到本地网络的任何计算机上打开浏览器，输入 http://192.168.4.4:8080/hello。连接到树莓派，Bottle 应该提供一个网页，包含"Hello Bottle World！"。只需几行代码，就可以创建一个 Web 服务器。

客户端将利用异步 JavaScript 和 XML（AJAX）框架，向服务器（运行在树莓派上的 Bottle）发出请求。为了让 AJAX 调用易于编写，我们将使用流行的 jQuery 库。

> **Python 装饰器**
>
> Python 中的装饰器采用@语法，它将一个函数作为参数，并返回另一个函数。装饰器提供了一种方便的方式，用另一个函数来"包装"一个函数。例如，这段代码：
>
> ```
> @wrapper
> def myFunc():
>     return 'hi'
> ```
>
> 相当于执行以下操作：
>
> ```
> myFunc = wrapper(myFunc)
> ```
>
> 函数是 Python 中的一等对象，可以像变量一样传递。

14.2.7 用 flot 绘制

现在让我们来看看如何绘制数据。flot 库有一个易于使用的强大 API，让我们用最少的代码来创建漂亮的图形。基本上，我们设置一些超文本标记语言（HTML）来保存一个图表，并提供值的一个数组来绘图，Flot 处理其余工作，如这个例子所示（可以在本书代码库的 simple-flot.html 文件中找到这段代码）。

```html
<html>
<head>
    <meta http-equiv="Content-Type" content="text/html; charset=utf-8">
    <title>SimpleFlot</title>
❶   <style>
        .demo-placeholder {
        width: 80%;
        height: 80%;
        }
    </style>
❷   <script language="javascript" type="text/javascript"
            src="flot/jquery.js"></script>
    <script language="javascript" type="text/javascript"
            src="flot/jquery.flot.js"></script>
    <script language="javascript" type="text/javascript">
❸   $(document).ready(function() {
        // create plot
❹       var data = [];
        for(var i = 0; i < 500; i ++) {
❺           data.push([i, Math.exp(-i/100)*Math.sin(Math.PI*i/10)]);
        }
❻       var plot = $.plot("#placeholder", [data]);
    });
    </script>
</head>

<body>
    <h3>A Simple Flot Plot</h3>
    <div class="demo-container">
❼       <div id="placeholder" class="demo-placeholder"></div>
    </div>
</body>
</html>
```

在❶行，定义一个 CSS 类（demo-placeholder），设置占位符元素的宽度和高度，来保存绘图（它将在文档的主体中定义）。在❷行，声明在此 HTML 文件中将使用的库的 JavaScript 文件：jquery.js 和 flot.js（请注意，jQuery 与 flot 捆绑在一起，因此不需要单独下载它。此外，请确保将顶层 flot 目录放在包含此项目所有源代码的同一目录中）。

接下来，用 JavaScript 生成要绘制的值。在❸行，用 jQuery 方法$(document).ready()定义一个函数，一旦 HTML 文件加载了，浏览器就执行它。在该函数内部，在❹行，声明一个空的 JavaScript 数组，然后循环 500 次，将形如[i, y]的值添加到该数组❺。每个值表示这个有趣函数（选择有点随意）的 x 和 y 坐标。

$$x(y) = e^{\frac{-x}{100}} \sin\left(\frac{2\pi}{10}x\right), \quad x \text{ 的范围是}[0, 500]$$

在❻行，从 flot 库调用 plot()来绘图。在❼行，plot()函数将包含绘图的 HTML 元素（placeholder 元素）的 id 作为输入。在浏览器中加载生成的 HTML 文件时，应该看到如图 14-4 所示的图形。

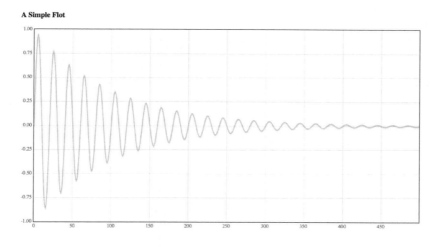

图 14-4　用 flot 创建的示例图形

这里使用了 flot 的默认设置，但你可以通过调整颜色，使用数据点而不是线条，添加图例和标题，采用交互式绘图，以及更多的方式来定制 flot 绘图（在 14.4.2 节，会看到设计天气数据的图表时如何操作）。

14.2.8　关闭树莓派

不要突然断开运行的树莓派的电源，否则可能会损坏文件系统，让树莓派无法启动。要关闭树莓派的用户界面（如果你直接或从计算机连接到该用户界面），请通过 SSH 输入以下内容：

```
$ sudo shutdown -h now
```

注意　　在运行上一个命令之前，需要确保已登录到你的树莓派。否则，你可能会关闭主计算机（如果你正在运行 Linux）。

输入 shutdown 命令几秒钟后，树莓派的黄色指示灯应该会闪烁 10 次。现在你可以安全地拔掉插头。

14.3　搭建硬件

除了前面提到的树莓派和外设之外，还需要以下各项和连接线：
- DHT11 传感器；
- 4.7kΩ 电阻；
- 100Ω 电阻；
- 红色 LED；
- 面包板。

图 14-5 展示了如何连接所有器件。DHT11 的 VDD 引脚连接到 +5V（树莓派上的引脚 #2），DHT11 的 DATA 引脚连接到树莓派上的引脚 #16，DHT11 的 GND 引脚连接到树莓派上的 GND（引脚 #6）。DATA 和 VDD 之间连接一个 4.7kΩ 电阻。LED 的阴极（负极）通过 100Ω 电阻器连接到 GND，阳极（正极）连接到树莓派的引脚 #18。

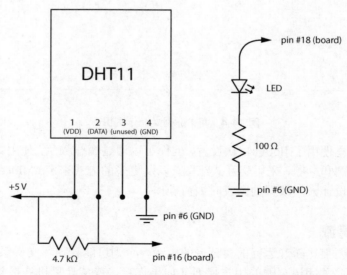

图 14-5　树莓派、DHT11 电路和 LED 之间的连接示意图

可以用无焊接的面包板连接 DHT11 和 LED 电路并测试该装置，如图 14-6 所示。当它的工作令人满意后，将该装置移到定制的外壳中。

图 14-6　树莓派、DHT11 和 LCD 与面包板连接

14.4　代码

现在，我们将开发在树莓派上运行的代码（如果想查看完整的项目代码，请跳到 14.5 节）。

下面是 main() 函数。

```
def main():
    print 'starting piweather...'
    # create parser
❶   parser = argparse.ArgumentParser(description="PiWeather...")
    # add expected arguments
    parser.add_argument('--ip', dest='ipAddr', required=True)
    parser.add_argument('--port', dest='portNum', required=True)

    # parse args
    args = parser.parse_args()

    # GPIO setup
❷   GPIO.setmode(GPIO.BOARD)
❸   GPIO.setup(18, GPIO.OUT)
❹   GPIO.output(18, False)
    # start server
❺   run(host=args.ipAddr, port=args.portNum, debug=True)
```

在❶行，设置了一个命令行参数解析器，它有两个必需参数的：--ip 表示服务器启动的 IP 地址，--port 表示服务器端口号。在❷行，启动 GPIO 引脚设置。我们使用 BOARD 模式，表明使用基于板的物理布局的引脚编号惯例。在❸行，将引脚 #18 设置为输出，因为我们打算向它写入数据来控制 LED，在❹行，设置引脚值为 False，所以 LED 在启动时关闭。然后通过提供 IP 地址和端口号启动 Bottle，并将 debug 设置为 True，监视所有警告消息❺。

14.4.1 处理传感器数据请求

现在快速看一下处理传感器数据请求的函数：

```
❶ @route('/getdata', method='GET')
❷ def getdata():
❸     RH, T = Adafruit_DHT.read_retry(Adafruit_DHT.DHT11, 23)
       # return dictionary
❹     return {"RH": RH, "T": T}
```

该方法定义了一个名为/getdata 的路由❶。客户端访问此路由定义的 URL /getdata 时，getdata()方法被调用❷，它利用 Adafruit_DHT 模块取得湿度和温度数据❸。在❹行，取得的数据作为字典返回，它将作为"JavaScript 对象表示法（JSON）"对象提供给客户端。JSON 对象包含由名—值对组成的一些列表，可以作为对象读入。在这个例子中，getdata()返回的 JSON 对象有个两名—值对：一个是湿度读数（RH），一个是温度（T）。

14.4.2 绘制数据

plot()函数处理客户端的绘图请求。该函数的第一部分定义了 HTML 的<head>部分，设置了"层叠样式表（CSS）"的样式，并加载了必要的 JavaScript 代码，如下所示：

```
  @route('/plot')
  def plot():
❶     return '''
  <html>
  <head>
      <meta http-equiv="Content-Type" content="text/html; charset=utf-8">
      <title>PiWeather</title>
      <style>
      .demo-placeholder {
      width: 90%;
      height: 50%;
      }
      </style>
          <script language="javascript" type="text/javascript"
      src="jquery.js"></script>
          <script language="javascript" type="text/javascript"
      src="jquery.flot.js"></script>
          <script language="javascript" type="text/javascript"
      src="jquery.flot.time.js"></script>
```

plot()函数是/plot URL 的 Bottle 路由，这表明连接到该 URL 时将调用 plot()方法。在❶行，plot()将整个 HTML 数据作为单个字符串返回，它将由客户端的 Web 浏览器显示。列表中的初始行是 HTML 标题，绘图的 CSS 大小声明和包含 flot 库的代码，所有这些都类似于生成图 14-4 中的 flot 绘图示例的设置。

代码的<body>元素显示了 HTML 的整体结构。

```
<body>
        <div id="header">
❶           <h2>Temperature/Humidity</h2>
        </div>

        <div id="content">
            <div class="demo-container">
❷               <div id="placeholder" class="demo-placeholder"></div>
            </div>
❸       <div id="ajax-panel"> </div>
        </div>
    <div>
❹       <input type="checkbox" id="ckLED" value="on">Enable Lighting.
❺       <span id="data-values"> </span>
    </div>

</body>
</html>
```

在❶行，只是为绘图添加了一个标题。在❷行添加了<placeholder>元素，稍后将由 flot 的 JavaScript 代码填充。在❸行，定义了一个 ID 为 ajax-panel 的 HTML 元素，它将显示所有 AJAX 错误，在❹行，创建了一个 ID 为 ckLED 的 checkbox 元素，它控制连接到树莓派的 LED。最后，创建了另一个 ID 为 data-values 的 HTML 元素❺，当用户单击绘图中的数据点时，可以显示传感器数据。JavaScript 代码将利用这些 ID 来访问和修改相应的元素。

现在让我们来深入了解嵌入的 JavaScript 代码，它们发起传感器数据请求，打开和关闭 LED。这段代码在<script language = "javascript"...>标记中，在 HTML 数据<head>下面。

```
❶ $(document).ready(function() {

    // plot options
❷   var options = {
        series: {
            lines: {show: true},
            points: {show: true}
            },
❸       grid: {clickable: true},
❹       yaxes: [{min: 0, max: 100}],
        xaxes: [{min: 0, max: 100}],
    };

    // create empty plot
❺   var plot = $.plot("#placeholder", [[]], options);
```

HTML 数据完全加载后，浏览器调用❶行的 ready()函数。在❷行，声明了一个 options 对象来定制绘图。在此对象中，可以让绘图显示线和点，在❸行，允许单击绘图网格（用于查询值）。在❹行，设置了坐标轴限制。在❺行，用 3 个参数调用 plot() 创建实际绘图：要在其中显示绘图的元素的 ID、值的数组（它开始为空），以及刚刚设置的 options 对象。

接下来，看看获取传感器数据的 JavaScript 代码。

```
    // initialize data arrays
❶   var RH = [];
    var T = [];
    var timeStamp = [];
    // get data from server
❷   function getData() {
        // AJAX callback
❸       function onDataReceived(jsonData) {
❹           timeStamp.push(Date());
            // add RH data
❺           RH.push(jsonData.RH);
            // removed oldest
❻           if (RH.length > 100) {
                RH.splice(0, 1);
            }
            // add T data
            T.push(jsonData.T);
            // removed oldest
            if (T.length > 100) {
                T.splice(0, 1);
            }
❼           s1 = [];
            s2 = [];
            for (var i = 0; i < RH.length; i++) {
                s1.push([i, RH[i]]);
                s2.push([i, T[i]]);
            }
            // set to plot
❽           plot.setData([s1, s2]);
            plot.draw();
        }

        // AJAX error handler
❾       function onError(){
            $('#ajax-panel').html('<p><strong>Ajax error!</strong> </p>');
        }

        // make the AJAX call
❿       $.ajax({
            url: "getdata",
            type: "GET",
            dataType: "json",
            success: onDataReceived,
            error: onError
        });
    }
```

在❶行，初始化温度和湿度值的空数组，以及一个 timeStamp 数组，保存收集每个值的时间。在❷行定义了 getData()方法，我们利用定时器定期调用该方法。在❸行，定义了 onDataReceived()方法，它被设置为 AJAX 调用的回调方法。在 JavaScript 中，可以在函数中定义函数，这些函数可以像常规变量一样使用，所以可以在 getData()函数中定义 onDataReceived()，然后将它作为回调函数传递给 AJAX 调用。

onDataReceived()函数创建了 JavaScript Date 对象，保存数据点的时间戳❹。来自服务器的数据利用 jsonData 对象传入 OnDataReceived()，在❺行，将来自此对象的温度数据记入数组。在❻行，如果元素数量超过 100，就删除数组中最早的元素，这将产生滚动图，类似第 12 章中为 Arduino 光传感器项目创建的滚动图。我们用相同的方式处理湿度数据。

在❼行，收集的数据被格式化，以便适合传递给 plot()方法。由于要同时绘制两个变量，所以需要一个包含 3 个图层的数组：

$$[[[i_0, RH_0], [i_1, RH_1], ...], [[i_0, T_0], [i_1, T_1]]$$

在❽行，数据被设置并绘制。

在❾行，定义了一个错误回调，AJAX 将使用之前设置的、ID 为 ajax-panel 的 HTML 元素来显示错误。在❿行，进行实际的 AJAX 调用，它指定 URL 为 getdata，即 Bottle 路由。该调用利用 HTTP 的 GET()方法，以 json 格式从服务器请求数据。AJAX 设置调用将 OnDataReceived()方法设置为成功回调，并将 onError()设置为错误回调。这个 AJAX 调用会立即返回，并在数据可用时，异步激活回调方法。

14.4.3 update()方法

现在看看每秒调用 getData()的 update()方法。

```
            // define an update function
            function update() {
                // get data
❶               getData();
                // set timeout
❷               setTimeout(update, 1000);
            }
            // call update
                update();
```

update()方法首先在❶行调用 getData()，然后在❷行利用 JavaScript 的 setTimeout() 方法在 1000 毫秒后调用它自己。因此，getData()每秒都会被调用，从服务器请求传感器数据。

14.4.4 用于 LED 的 JavaScript 处理程序

现在来看看 LED 复选框的 JavaScript 处理程序，它向 Web 服务器发送一个 AJAX 请求。

```
        // define the click handler for the LED control button
❶   $('#ckLED').click(function() {
❷       var isChecked = $("#ckLED").is(":checked") ? 1:0;
❸       $.ajax({
            url: '/ledctrl',
            type: 'POST',
            data: { strID:'ckLED', strState:isChecked }
        });
    });
```

在❶行，为前面在 HTML 中创建的 ID 为 ckLED 的复选框定义了一个点击处理程序。每次用户单击复选框时，将调用这个单击处理程序函数。该函数保存复选框的状态❷，然后❸行的代码利用 URL /ledctrl 和 HTTP 请求类型 POST 来调用 AJAX，它将选中的状态作为数据发送。

这个 AJAX 请求的服务器端处理程序根据请求将树莓派的 GPIO 引脚设置为打开或关闭。

```
❶   @route('/ledctrl', method='POST')
    def ledctrl():
❷       val = request.forms.get('strState')
❸       on = bool(int(val))
❹       GPIO.output(18, on)
```

在❶行，定义了 URL /ledctrl 的 Bottle 路由，这是处理此请求的 ledctrl() 方法的装饰器。在❷行，利用 Bottle 的 request 对象访问由客户端代码发送的 strState 参数的字符串值，该值在❸行转换为布尔值。对引脚#18 使用 GPIO.output() 方法，打开或关闭连接到此引脚的 LED❹。

14.4.5 添加交互性

我们还希望为绘图提供一些交互性。flot 库提供了一种方式让用户点击数据点获取值，这是在调用 plot() 函数时，通过在 options 中传入 clickable:true 而启用的。下面，我们定义在单击数据点时要调用的函数：

```
    $("#placeholder").bind("plotclick", function (event, pos, item) {
        if (item) {
❶           plot.highlight(item.series, item.datapoint);
❷           var strData = ' [Clicked Data: ' +
                        timeStamp[item.dataIndex] + ': T = ' +
                        T[item.dataIndex] + ', RH = ' + RH[item.dataIndex]
                        + ']';
❸           $('#data-values').html(strData);
        }
    });
});

</script>
</head>
```

该函数在❶行调用 flot 的 highlight()方法，在单击的点周围绘制一个圆环。在❷行，它准备一个字符串，以显示在 ID 为 data-value 的 HTML 元素中。点击数据点时，flot 会向函数传入一个 item 对象，它有一个名为 dataIndex 的成员。我们可以利用此索引，从 ready()函数中定义的时间戳、温度和湿度数组中取得相关数据。最后，将字符串添加到 HTML 元素❸。

这就是 Python 代码中嵌入的 JavaScript，但我们还需要一个 Bottle 路由来找到 Web 页面需要的 JavaScript 文件，如下所示：

```
@route('/<filename:re:.*\.js>')
def javascripts(filename):
    return static_file(filename, root='flot')
```

这段代码告诉 Bottle 服务器在子目录 float/下查找这些文件，它在与程序同级的目录下。

14.5　完整代码

可以在 https://github.com/electronut/pp/tree/master/piweather/piweather.py 找到此项目的完整代码列表。

```
from bottle import route, run, request, response
from bottle import static_file
import random, argparse
import RPi.GPIO as GPIO
from time import sleep
import Adafruit_DHT

@route('/hello')
def hello():
    return "Hello Bottle World!"

@route('/<filename:re:.*\.js>')
def javascripts(filename):
    return static_file(filename, root='flot')

@route('/plot')
def plot():
    return '''
<html>
<head>
    <meta http-equiv="Content-Type" content="text/html; charset=utf-8">
    <title>PiWeather</title>
    <style>
    .demo-placeholder {
    width: 90%;
    height: 50%;
    }
    </style>
    <script language="javascript" type="text/javascript"
        src="jquery.js"></script>
    <script language="javascript" type="text/javascript"
```

```
            src="jquery.flot.js"></script>
    <script language="javascript" type="text/javascript"
            src="jquery.flot.time.js"></script>
    <script language="javascript" type="text/javascript">

$(document).ready(function() {

    // plot options
    var options = {
        series: {
            lines: {show: true},
            points: {show: true}
              },
        grid: {clickable: true},
        yaxes: [{min: 0, max: 100}],
        xaxes: [{min: 0, max: 100}],
    };

    // create empty plot
    var plot = $.plot("#placeholder", [[]], options);

    // initialize data arrays
    var RH = [];
    var T = [];
    var timeStamp = [];
    // get data from server
    function getData() {
        // AJAX callback
        function onDataReceived(jsonData) {
            timeStamp.push(Date());
        // add RH data
        RH.push(jsonData.RH);
        // removed oldest
        if (RH.length > 100) {
            RH.splice(0, 1);
        }
        // add T data
        T.push(jsonData.T);
        // removed oldest
        if (T.length > 100) {
            T.splice(0, 1);
        }
        s1 = [];
        s2 = [];
        for (var i = 0; i < RH.length; i++) {
            s1.push([i, RH[i]]);
            s2.push([i, T[i]]);
        }
        // set to plot
        plot.setData([s1, s2]);
        plot.draw();
    }

    // AJAX error handler
    function onError(){
    $('#ajax-panel').html('<p><strong>Ajax error!</strong> </p>');
    }
```

```
        // make the AJAX call
        $.ajax({
            url: "getdata",
            type: "GET",
            dataType: "json",
            success: onDataReceived,
            error: onError
        });
    }

    // define an update function
    function update() {
        // get data
        getData();
        // set timeout
        setTimeout(update, 1000);
    }

    // call update
    update();

    // define click handler for LED control button
    $('#ckLED').click(function() {
        var isChecked = $("#ckLED").is(":checked") ? 1:0;
        $.ajax({
            url: '/ledctrl',
            type: 'POST',
                data: { strID:'ckLED', strState:isChecked }
            });
        });

        $("#placeholder").bind("plotclick", function (event, pos, item) {
            if (item) {
                plot.highlight(item.series, item.datapoint);
                var strData = ' [Clicked Data: ' +
                              timeStamp[item.dataIndex] + ': T = ' +
                              T[item.dataIndex] + ', RH = ' + RH[item.dataIndex]
                              + ']';
                $('#data-values').html(strData);
            }
        });
});

</script>
</head>

<body>
    <div id="header">
        <h2>Temperature/Humidity</h2>
    </div>

    <div id="content">
        <div class="demo-container">
            <div id="placeholder" class="demo-placeholder"></div>
        </div>
        <div id="ajax-panel"> </div>
    </div>
    <div>
```

```
            <input type="checkbox" id="ckLED" value="on">Enable Lighting.
            <span id="data-values"> </span>
        </div>

</body>
</html>
'''

@route('/getdata', method='GET')
def getdata():
    RH, T = Adafruit_DHT.read_retry(Adafruit_DHT.DHT11, 23)
    # return dictionary
    return {"RH": RH, "T": T}

@route('/ledctrl', method='POST')
    def ledctrl():
    val = request.forms.get('strState')
    on = bool(int(val))
    GPIO.output(18, on)

# main() function
def main():
    print 'starting piweather...'
    # create parser
    parser = argparse.ArgumentParser(description="PiWeather...")
    # add expected arguments
    parser.add_argument('--ip', dest='ipAddr', required=True)
    parser.add_argument('--port', dest='portNum', required=True)

    # parse args
    args = parser.parse_args()

    # GPIO setup
    GPIO.setmode(GPIO.BOARD)
    GPIO.setup(18, GPIO.OUT)
    GPIO.output(18, False)
    # start server
    run(host=args.ipAddr, port=args.portNum, debug=True)

# call main
if __name__ == '__main__':
    main()
```

14.6 运行程序

将树莓派连接到 DHT11 和 LCD 电路后,利用 SSH 从计算机登录树莓派,并输入以下内容(替换你为树莓派设置的 IP 地址和端口):

```
$ sudo python piweather.py --ip 192.168.x.x --port xxx
```

现在打开浏览器,在浏览器的地址栏中输入树莓派的 IP 地址和端口,形式为:

```
http://192.168.x.x:port/plot
```

你应该看到如图 14-7 所示的图形。

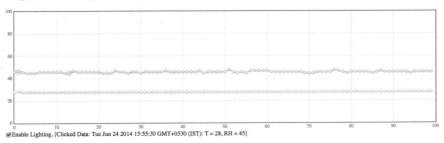

图 14-7　piweather.py 示例运行的结果网页和图表

如果点击图表中的任何数据点，点击的数据部分应该更新，显示有关该点的更多信息。在收集到 100 个点后，当新数据进入时，图形将开始水平滚动（单击 Enable Lighting 复选框来打开和关闭 LED）。

14.7　小结

本项目构建了一个基于树莓派的天气监视器，通过 Web 界面绘制温度和湿度数据。下面是本项目涵盖的一些概念：
- 设置树莓派；
- 使用树莓派的 GPIO 引脚与硬件通信；
- 与 DHT11 温湿度传感器接口；
- 使用 Python Web 框架 Bottle 来创建 Web 服务器；
- 使用 JavaScript 库 flot 来制作图表；
- 构建客户端—服务器应用程序；
- 通过 Web 界面控制硬件。

14.8　实验

请尝试通过以下修改来改进此项目。

1. 提供一种导出传感器数据的方法。一种简单的方法是在服务器上维护一个（T，RH）元组的列表，然后为/export 写一个 Bottle 路由，并在这个方法中返回 CSV 格式的值。修改/plot 路由，使其包含 HTML 代码，在网页上放置"Export"按钮。单击该按钮时，有相应的 AJAX 代码调用服务器上的 export()方法，它将发送要在浏览器中显示的 CSV 值。然后，这些值可以让用户从浏览器窗口中复制或保存。

2. 该程序绘制 DHT11 数据，但 100 个值后，它开始滚动。如果想查看更长时间的历史数据怎么办？一种方法是在服务器上维护一个较长的（T，RH）元组列表，然后修改服务器代码，利用一个按钮来发送 HTML 数据，并使用必要的 JavaScript 代码，在绘制完整范围的数据或仅绘制最近 100 个值之间切换。如何取得树莓派关闭之前的老数据（提示：当服务器退出时，将数据写入文本文件，并在启动时加载老数据）？要使此项目真正可扩展，请使用数据库，如 SQLite。

附录 A

软件安装

在本附录中,我将介绍如何安装 Python,以及本书中使用的外部模块和代码。由于第 14 章中已经介绍了几个树莓派相关项目的安装,这里将跳过这些说明。本书中的项目已经用 Python 2.7.8 和 Python 3.3.3 进行了测试。

A.1 安装本书项目的源代码

你可以从 https://github.com/electronut/pp/ 下载这本书的项目的源代码。使用此网站上的 Download ZIP 选项来获取代码。

下载并解压缩代码后,需要将下载的代码中的 common 文件夹的路径(通常为 pp-master/common)添加到 PYTHONPATH 环境变量中,以便模块可以查找和使用这些 Python 文件。

在 Windows 上，实现的方式是创建 PYTHONPATH 环境变量，如果它已经存在，则添加它。在 OS X 上，可以将此行添加到主目录中的 .profile 文件（如果需要就创建一个文件）：

```
export PYTHONPATH=$PYTHONPATH:path_to_common_folder
```

Linux 用户可以执行类似于 OS X 的操作，在 .bashrc、.bash_profile 或 .cshrc/.login 中添加。可以使用 echo $SHELL 命令查看默认 shell。

现在，来看看如何在 Windows、OS X 和 Linux 上安装 Python 以及本书中使用的模块。

A.2 在 Windows 上安装

首先，从 https://www.python.org/download/ 下载并安装 Python。

A.2.1 安装 GLFW

对于本书中基于 OpenGL 的 3D 图形项目，需要 GLFW 库，你可以从 http://www.glfw.org/download.html 下载。

在 Windows 上，安装 GLFW 后，将 GLFW_LIBRARY 环境变量（在搜索栏中键入 Edit Environment Variables）设置为安装 glfw3.dll 的完整路径，以便 GLFW 的 Python 绑定能找到此库。路径看起来类似 C:\ glfw-3.0.4.bin.WIN32\lib-msvc120\glfw3.dll。

在 Python 中使用 GLFW，要用一个名为 pyglfw 的模块，它由一个 Python 文件组成，名为 glfw.py。你不需要安装 pyglfw，因为本书的源代码已包含它，可以在 common 目录中找到它。但是，如果需要安装较新的版本，来源是 https://github.com/rougier/pyglfw/。

你还需要确保计算机已安装显卡驱动程序。这通常是有好处的，因为许多软件程序（特别是游戏）都会利用图形处理单元（GPU）。

A.2.2 为每个模块安装预先构建二进制文件

在 Windows 上安装必要的 Python 模块，最简单的方法是获取预先构建的二进制文件。每个模块的链接在下面列出。请下载适当的安装程序（32 或 64 位）。根据 Windows 设置，你可能需要以管理员权限来运行这些安装程序。

pyaudio

http://www.lfd.uci.edu/~gohlke/pythonlibs/#pyaudio

pyserial

http://www.lfd.uci.edu/~gohlke/pythonlibs/#pyserial

scipy

http://www.lfd.uci.edu/~gohlke/pythonlibs/#scipy

http://sourceforge.net/projects/scipy/files/scipy/

numpy

http://www.lfd.uci.edu/~gohlke/pythonlibs/#numpy

http://sourceforge.net/projects/numpy/files/NumPy/

pygame

http://www.lfd.uci.edu/~gohlke/pythonlibs/#pygame

Pillow

http://www.lfd.uci.edu/~gohlke/pythonlibs/#pillow

https://pypi.python.org/pypi/Pillow/2.5.0#downloads

pyopengl

http://www.lfd.uci.edu/~gohlke/pythonlibs/#pyopengl

matplotlib

http://www.lfd.uci.edu/~gohlke/pythonlibs/#matplotlib

matplotlib 库依赖于 dateutil、pytz、pyparsing 和 six，这些依赖库可以从以下链接获得：

dateutil

http://www.lfd.uci.edu/~gohlke/pythonlibs/#python-dateutil

pytz

http://www.lfd.uci.edu/~gohlke/pythonlibs/#pytz

pyparsing

http://www.lfd.uci.edu/~gohlke/pythonlibs/#pyparsing

six

http://www.lfd.uci.edu/~gohlke/pythonlibs/#six

A.2.3 其他选项

你也可以安装适当的编译器,在 Windows 上自己构建所有必需的软件包。有关兼容编译器的列表,参见 https://docs.python.org/2/install/index.html#gnu-c-cygwin-mingw。另一个选择是在 http://www.scipy.org/install.html 安装特殊的 Python 发行版,其中预装了大部分这些软件包。

A.3 在 OS X 上安装

以下是在 OS X 上安装 Python 和必要模块的建议步骤。

A.3.1 安装 Xcode 和 MacPorts

第一步是安装 Xcode。你可以通过 App Store 获取它。如果你运行的是旧版本的操作系统,则可以从 Apple 开发人员网站(https://developer.apple.com/)获取兼容版本的 Xcode。安装 Xcode 后,请确保也安装了命令行工具。下一步是安装 MacPorts。你可以参考 MacPorts 指南(http://guide.macports.org/#installing.xcode),其中包含了详细的安装说明,以帮助你完成此过程。

MacPorts 安装自己的 Python 版本,最简单的方法是你的项目就使用该版本(OS X 也内置了 Python,但在它上面安装模块有许多问题,所以最好不要管它)。

A.3.2 安装模块

安装 MacPorts 后,可以使用 Terminal(终端)应用程序中的 port 命令,安装本书所需的模块。

在终端窗口中使用下面的命令检查 Python 的版本:

```
$ port select --list python
```

如果你有多个 Python 安装,可以使用下面的命令,选择特定版本的 Python 作为 MacPorts 的活动版本(这里选择了 Python 版本 2.7):

```
$ port select --set python python27
```

然后可以安装所需的模块。在终端窗口中逐个运行这些命令。

```
sudo port install py27-numpy
sudo port install py27-Pillow
```

```
sudo port install py27-matplotlib
sudo port install py27-opengl
sudo port install glfw
sudo port install py27-scipy
sudo port install py27-pyaudio
sudo port install py27-serial
sudo port install py27-game
```

MacPorts 通常将 Python 安装在/opt/local/中。可以通过在.profile 中设置 PATH 环境变量，来确保在终端窗口中获得正确的 Python 版本。这是我的设置：

```
PATH=/opt/local/Library/Frameworks/Python.framework/Versions/2.7/bin:$PATH
export PATH
```

这段代码确保正确的 Python 版本可以从任何终端运行。

A.4 在 Linux 上安装

Linux 通常内置 Python 并且构建所需软件包的所有开发工具。在大多数 Linux 发行版上，应该能够使用 pip 获取本书所需的包。有关安装 pip 的说明，请参阅以下链接：http://pip.readthedocs.org/en/latest/installing.html。

可以用 pip 安装一个包，像这样：

```
sudo pip install matplotlib
```

另一种安装软件包的方法是下载模块源代码，通常是.gz 或.zip 文件。将这些文件解压缩到文件夹后，可以按如下方法安装它们：

```
sudo python setup.py install
```

对于本书所需的每个包，你需要使用这些方法中的一种。

附录 B
基础实用电子学

本附录简要介绍了搭建电子电路相关的一些基本术语、组件和工具。电子学是工程学分支,涉及利用有源和无源电子元件,设计和构造电路。

这个主题很大,所以我仅仅论及皮毛[①]。但从爱好者或 DIY(自己动手)的角度来看,你不需要知道一大堆知识,就能开始玩电子电路。你可以在搭建中学习,根据我自己的经验,这是一个有趣的、令人上瘾的爱好。我希望通过快速介绍以下主题,激励你开始阅读这方面的内容,并让你开始设计和搭建自己的电路。

B.1 常用组件

大多数电子部件使用半导体类的材料。半导体具有特殊的电气特性,使其成为电子器件构造的理想选择。

---

① 关于电子学的全面参考,我推荐 Paul Scherz 和 Simon Monk 的《Practical Electronics for Inventors》第 3 版(McGraw-Hill,2013)。

在本节中，我们将介绍电路中使用的一些最常见的组件。图 B-1 展示了这些组件，以及在电路图中表示它们的符号。

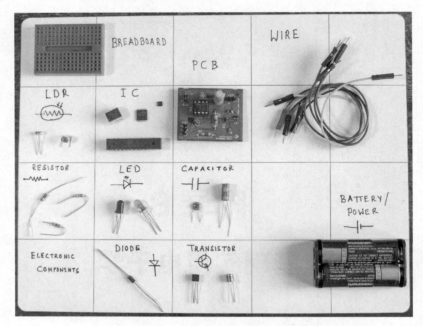

图 B-1　常用电子组件及其相应的符号

B.1.1　面包板

面包板是用于原型电子电路的多孔模块。面包板的孔中有带弹簧的夹子，并且以简单的方式互连，容易进行实验。不必焊接每个连接，只要将组件插入面包板，并用电线连接它们。

B.1.2　光敏电阻（LDR）

LDR 是一种电阻器，其电阻随着照在其上的光的强度而减小。它用作电子电路中的光传感器。

B.1.3　集成电路（IC）

IC 是包含完整电子电路的器件。IC 非常小，每平方厘米可以包含数十亿晶体管。每个 IC 通常都有一种特定的应用，制造商的数据表提供了必要的原理图、电气和物理特性，以及示例应用程序，以帮助用户。你可能会遇到的常见 IC 是 555，它的主要用作定时器。

B.1.4 印刷电路板（PCB）

要制作电子电路，需要一个地方来组装组件。通常在 PCB 上完成此操作，该 PCB 由绝缘体组成，绝缘体上有一层或多层导电材料（通常为铜）。导电层成形的方式让它形成电路的布线。

元件作为"通孔"元件或"表面安装"元件安装到 PCB 上，通过焊接与导电层实现导电连接。

B.1.5 电线

搭电路不能没有电线。我们将使用塑料绝缘的铜线。

B.1.6 电阻

电阻是电路中最常见的元件之一。电阻用于减小电路中的电流或电压，并且根据其阻值（以欧姆为单位）来指定。例如，2.7k 欧姆电阻具有 2.7 千欧姆或 2700 欧姆的阻值。电阻具有指示其阻值的彩色编码带，以及两个可互换的引线，因为它没有极性。

B.1.7 发光二极管（LED）

LED 是在许多电路中看到的小闪烁的灯。然而，LED 是特殊类型的二极管，因此它也具有极性，并需要根据极性连接。它通常与电阻器结合使用，限制流过它的电流，因此不会损坏。不同颜色的 LED 具有不同的最小"导通"电压。

B.1.8 电容器

电容是有两个引线的器件，用于存储电荷。它是根据电容量来测量的，其单位为法拉。典型电容的电容量以微法（μF）为单位。电容有极化和非极化两种。

B.1.9 二极管

二极管是允许电流仅在一个方向上通过的电子器件。二极管通常用作整流器（将 AC 电流转换为 DC 的器件）。二极管有两根引线：阳极和阴极。这意味着它有极性，所以两根引线需要与电路的其余部分正确连接。

B.1.10 晶体管

晶体管可以看成是电子开关。晶体管还可以用作电流或电压的放大器。作为集成电路的基本构件，它是最重要的电子部件之一。晶体管有几种类型，但最常见的是双极型晶体管（BJT）和金属氧化物半导体场效应晶体管（MOSFET）。晶体管通常有三根引线。对于 BJT，它们被命名为基极、集电极和发射极，而对于 MOSFET，它们被命名为栅极、源极和漏极。当晶体管用作开关时，BJT 的基极电流（或 MOSFET 的栅极电压），是让电流能够流过集电极和发射极的原因，因此它作为开关，控制外部负载，如 LED 或继电器。

B.1.11 电池/电源

大多数电子设备在 3 到 9 伏的小电压下工作，这种电压可以由电池提供，或利用插入墙上交流电源插座的电源适配器提供。

B.2 基本工具

除了刚才描述的组件，你还需要一些基本工具，才能开始搭建电子电路。图 B-2 展示了业余爱好者的工作台上通常会看到的工具。接下来介绍其中一些基本工具。

图 B-2　典型的电子工作台，有万用表、工作灯、夹具、剥线钳、螺丝刀、尖嘴钳、放大镜、焊锡、焊剂、示波器和焊接台

B.2.1 万用表

万用表是用于测量电路的电气特性（例如电压、电流、电容和电阻）的仪器。它还经常用于测量电路中的连续性，这表示电流的连续流动。当你试图调试电路时，万用表是非常有用的。

B.2.2 烙铁及配件

对电路在面包板上工作感到满意后，下一步就是将它转移到 PCB 上，这需要焊接。焊接是使用加热的填充金属来接合两种金属的过程。填料或焊料，过去含铅，但现在常用无铅焊料合金，这更环保。要焊接元件时，将元件放在 PCB 上，加上助焊剂（使焊接过程更容易的化学品），然后将热的焊料与烙铁一起使用。焊料冷却时，它在部件和铜箔层之间形成物理结合和电连接。

B.2.3 示波器

示波器是用于测量和显示来自电子电路的电压的仪器。它是分析电压波形的有用工具。例如，可以用它来调试从传感器输出的数字数据，或测量从音频放大器输出的模拟电压。它还有其他专门的测量功能，如快速傅里叶变换（FFT）和均方根（RMS）。

图 B-2 还展示了搭建电路的一些其他有用的工具：多头螺丝刀、尖嘴钳、剥线钳、用于在焊接时固定 PCB 的夹具、让工作区域照明良好的台灯、帮助你查看小元件和焊接点的放大镜，以及用于烙铁头的清洁器。

B.3 搭建电路

在搭建电路时，从电路图或原理图开始，它将告诉你组件如何相互连接。通常，你会在面包板上搭建该电路，插入元件并用导线连接它们。测试电路并满意后，你可能会考虑把它移到 PCB 上。虽然面包板很方便，但它看起来有点乱糟糟，这些松散的电线在部署时会不可靠。

你可以使用通用 PCB，它的底部有固定的铜图案，也可以设计自己的 PCB。前者很适合小电路。通常，你会焊接元件，利用已经存在的铜箔连接，然后根据需要通过焊接额外的电线来搞定其余部分。图 B-3 展示了一个简单电路的原理图、面包板原型和 PCB 结构。

如果你想要非常漂亮的 PCB，可以自己设计一个并制造出来，也花费不多。有几个软件包可用于设计 PCB，但最常用的（免费）有 EAGLE[1] 和 KiCad[2]。EAGLE 软件有一个简版是免费的，条件是只用于非营利应用。它有一定的限制（例如，只有两个铜层，最大 PCB 尺寸为 10 厘米×8 厘米，每个项目一张原理图），但这些对

---

[1] CadSoft EAGLE PCB 设计软件，http://www.cadsoftusa.com/eagle-pcb-design-software/。
[2] KiCad EDA 软件套件，http://www.kicad-pcb.org/display/KICAD/KiCad+EDA+Software+Suite/。

于爱好者来说应该不会有太大的问题。但是，EAGLE 的学习曲线有些陡峭，而且我觉得界面有点令人困惑。在开始之前，我建议你先看一些视频教程[①]。

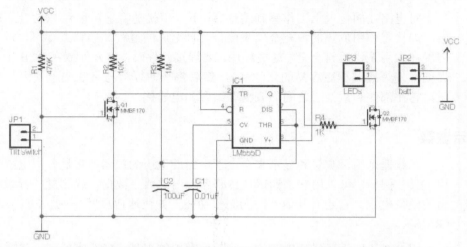

图 B-3　从原理图到面包板原型，再到 PCB 电路

  EAGLE 中的典型工作流程如下：首先，在原理图编辑器中创建电路图。为此，需要添加电路中使用的元件，然后连线。EAGLE 有大量的元件库，很可能你的组件已经在那些库中列出（EAGLE 还允许你创建自己定义的组件）。完成原理图后，EAGLE 就可以从中生成一个 PCB，从而生成这些元件的物理表示。然后，需要放置元件并完成电路布线，设计 PCB 上铜箔层的连接路径。典型的 EAGLE 设计如图 B-4 所示。左边是原理图，右边是相应的电路板。使用 EAGLE 需要一些练习，但 YouTube 教程将帮助你起步。

---

① Jeremy Blum, "Tutorial 1 for Eagle: Schematic Design", YouTube（2012 年 6 月 14 日），https://www.youtube.com/watch?v=1AXwjZoyNno。

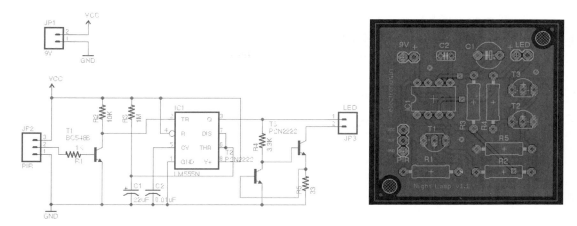

图 B-4　使用 EAGLE 创建的电路原理图和相应的 PCB 设计

设计了 PCB 后，你需要制造它。一些制造 PCB 的技术可以在家里完成[1]，这可以很有趣，但更专业的技术是将你的设计发送给 PCB 制造商。这些公司通常接受 Gerber 设计格式，你可以直接从 EAGLE 生成这些文件，只需一点设置[2]。许多公司制造 PCB。我接触过一个很好的公司是 OSH Park[3]。

完成了 PCB 和元件焊接后，要考虑你的项目的外壳。当前的制造技术允许你用激光打印和 3D 打印等技术，设计并构建专业外观封装。你可以使用 2D[4]和 3D[5]软件的组合来设计你的作品。Rich Decibels 的一个项目很好地说明了这里讨论的整个构建过程[6]。

B.4　下一步

你可以从两个角度来探索实用电子学。第一个是从头开始，学习将简单的电路放在一起，了解模拟和数字电路，最后继续了解微控制器和与计算机的接口电路。另一种方法是从编程的角度开始，即从 Arduino 和树莓派等硬件友好的板卡开始，然后通过为这些板卡搭建传感器和致动器来了解电路。两种方法都很好，大多数时候你可能会发现自己处于两者之间。

---

[1] 在家里制作 PCB 的一些技术可以在 https://embeddedinn.wordpress.com/tutorials/home-made-sigle-sided-pcbs/ 找到。
[2] Akash Patel,"Generating Gerber files from EAGLE",YouTube(2010 年 4 月 7 日),https://www.youtube.com/watch?v=B_SbQeF83XU。
[3] https://oshpark.com/。
[4] 对于 2D 软件，请参阅 Inkscape，http://www.inkscape.org/en/。
[5] 对于 3D 软件，请参阅 SketchUp，http://www.sketchup.com/。
[6] Rich Decibels,"Laser-Cut Project Box Tutorial,"Ponoko(2011 年 8 月 9 日),http://support.ponoko.com/entries/20344437-Laser-cut-project-box-tutorial/。

预祝你在电子工程方面取得巨大成功。一旦你创建了某个作品，我希望你花时间记录下来，并与世界分享。在 Instructables（http://www.instructables.com/）这样的网站上，可以分享你的项目，或从全球其他人的 DIY 项目中寻找灵感。

附录 C
树莓派的建议和技巧

正如你在第 14 章中学到的,树莓派是一个带有操作系统的完整计算机,这意味着需要一些设置过程才能开始使用它。在该项目中,我介绍了基本设置,但这里有一些额外的建议和技巧,为项目准备树莓派。

C.1 设置 Wi-Fi

在第 14 章中,我建议使用内置的 Wi-Fi 配置实用程序,在树莓派上设置 Wi-Fi。但利用命令行,可以更快地做同样的事。首先,在终端中输入以下内容,在 nano 编辑器中显示配置文件:

```
$ sudo nano /etc/wpa_supplicant/wpa_supplicant.conf
```

我的文件看起来像这样：

```
ctrl_interface=DIR=/var/run/wpa_supplicant GROUP=netdev
update_config=1

network={
    ssid=your-WiFi-network-name
    psk=your-password
    proto=RSN
    key_mgmt=WPA-PSK
    pairwise=TKIP
    auth_alg=OPEN
}
```

编辑文件的内容，设置 ssid 和 psk 以符合你的 Wi-Fi 设置。如果缺少整个 network 部分，请输入设置（填写与你自己的 Wi-Fi 设置相符的详细信息）。

C.2 检查树莓派是否已连接

可以用另一台计算机的 ping 命令，检查树莓派是否连接到本地网络。下面是 ping 会话看起来的样子：

```
$ ping 192.168.4.32
PING 192.168.4.32 (192.168.4.32): 56 data bytes
64 bytes from 192.168.4.32: icmp_seq=0 ttl=64 time=13.677 ms
64 bytes from 192.168.4.32: icmp_seq=1 ttl=64 time=8.277 ms
64 bytes from 192.168.4.32: icmp_seq=2 ttl=64 time=9.313 ms
--snip--
```

这个 ping 的输出显示了发送的字节数和获取应答所花费的时间。如果你看到消息 "Request timeout..."，就知道你的树莓派未连接到网络。

C.3 防止 Wi-Fi 适配器进入睡眠状态

如果在一段时间后无法 ping 通树莓派或用 SHH 连接，你的 USB Wi-Fi 适配器可能已进入睡眠状态。可以通过禁用电源管理来防止这种情况发生。首先，使用下面的命令打开控制电源管理的文件：

```
$ sudo nano /etc/modprobe.d/8192cu.conf
```

然后将以下内容添加到该文件中：

```
# disable power management
options 8192cu rtw_power_mgnt=0
```

重新启动树莓派，Wi-Fi 适配器现在应该一直保持工作。

C.4　从树莓派备份代码和数据

当你在树莓派上放置越来越多的代码时，需要一种备份文件的方法。rsync 是一个了不起的工具，它可以同步两个文件夹之间的文件，即使它们存在于网络上的不同机器上。rsync 非常强大，所以不要乱动它，除非很小心，否则可能会导致删除原始文件。如果你是第一次用它，请首先备份你的测试文件，并使用-n 标志（或者叫"彩排"）。这样，rsync 只告诉你它会做什么，而不实际做，这样你有机会熟悉该程序，不会意外删除任何东西。下面是我的脚本，从树莓派（递归地）备份代码目录到运行 OS X 的计算机上：

```
#!/bin/bash
echo Backing up RPi \#1...

# set this to Raspberry Pi IP address
PI_ADDR="192.168.4.31"

# set this to the Raspberry Pi code directory
# note that the trailing slash (/) is important with rsync
PI_DIR="code/"

# set this to the local code (backup) directory
BKUP_DIR="/Users/mahesh/code/rpi1/"

# run rsync
# use this first to test:
# rsync -uvrn pi@$PI_ADDR:$PI_DIR $BKUP_DIR
rsync -uvr pi@$PI_ADDR:$PI_DIR $BKUP_DIR

echo ...
echo done.

# play sound (for OS X only)
afplay /System/Library/Sounds/Basso.aiff
```

请修改这段代码，以符合你的目录。对于 Linux 和 OS X 用户，rsync 已内置。Windows 用户可以尝试 grsync[①]。

C.5　备份整个树莓派操作系统

备份树莓派上的操作系统是一个好主意。这样，如果由于不正常关机导致 SD 卡上的文件系统损坏，你可以快速地将操作系统重写到 SD 卡，而无需重新完成整

---

① 你可以从 http://grsync-win.sourceforge.net/ 下载 grsync - 用于 Windows 的 rsync 端口。

个设置。你还可以利用备份将现有安装克隆到另一个树莓派上。StackExchange 上发布的一个解决方案[1]介绍了在 Linux 和 OS X 上使用 dd 工具，或在 Windows 上使用 Win32 Disk Imager 软件。

C.6 利用 SSH 登录到树莓派

在第 14 章中，我讨论了如何方便地利用 SSH 登录到树莓派并工作。如果你不仅频繁地这样做，而且是用同一台计算机，可能会觉得每次输入密码很烦人。使用 SSH 附带的 ssh-keygen 实用程序，你可以设置公共/私有密钥方案，以便安全地登录到树莓派，而无需输入密码。对于 OS X 和 Linux 用户，请按照下列步骤操作（对于 Windows 用户，PuTTY 允许你进行类似操作[2]）。在计算机的终端窗口输入以下内容，根据需要修改树莓派的 IP 地址：

```
$ ssh-keygen
Generating public/private rsa key pair.
Enter file in which to save the key (/Users/xxx/.ssh/id_rsa):
Enter passphrase (empty for no passphrase):
Enter same passphrase again:
Your identification has been saved in /Users/xxx/.ssh/id_rsa.
Your public key has been saved in /Users/xxx/.ssh/id_rsa.pub.
The key fingerprint is:
--snip--
```

现在，将此密钥文件复制到树莓派：

```
$ scp ~/.ssh/id_rsa.pub pi@192.168.4.32:.ssh/
The authenticity of host '192.168.4.32 (192.168.4.32)' can't be established.
RSA key fingerprint is f1:ab:07:e7:dc:2e:f1:37:1b:6f:9b:66:85:2a:33:a7.
Are you sure you want to continue connecting (yes/no)? yes
Warning: Permanently added '192.168.4.32' (RSA) to the list of known hosts.
pi@192.168.4.32's password:
id_rsa.pub                                      100%  398     0.4KB/s   00:00
```

然后，登录到树莓派：

```
$ ssh pi@192.168.4.32
pi@192.168.4.32's password:

$ cd .ssh
$ ls
id_rsa.pub known_hosts
```

---

[1] 你可以在 Stack Exchange 上找到备份树莓派的 SD 卡的说明。"How do I backup my Raspberry Pi?"，Stack Exchange，http://raspberrypi.stackexchange.com/questions/311/how-do-i-backup-my-raspberry-pi/。

[2] "How to Create SSH Keys with PuTTY to Connect to a VPS"，DigitalOcean（2013 年 7 月 19 日），https://www.digitalocean.com/community/tutorials/how-to-create-ssh-keys-with -putty-to-connect-to-a-vps/。

```
$ cat id_rsa.pub >> authorized_keys
$ ls
authorized_keys id_rsa.pub known_hosts
$ logout
```

下次登录树莓派时，系统不会要求你输入密码。另外，请注意，我在 ssh-keygen 中使用一个空密码，这是不安全的。这个设置可能适用于树莓派硬件项目，这里你不太在意安全性，但有关使用 SSH 密码的进一步讨论，请参阅"Working with SSH Key Passphrases"，这是一篇有用的 GitHub 文章①。

C.7　使用树莓派相机

如果你想用树莓派拍摄照片，有一个专用的相机模块②。该模块有一个固定焦距和景深的相机，相机支持图像记录（5 百万像素）和视频记录（1080 像素在 30 帧/秒）。它通过带状电缆连接到树莓派。安装相机模块后，可以使用 raspistill 命令拍摄照片或视频。请确保在首次启动树莓派时启用相机支持。在安装摄像机之前，请查看安装视频③。这段视频还介绍了如何使用 raspistill 命令。④

C.8　在树莓派上启用声音

树莓派带有一个音频输出插孔。如果你无法在树莓派上听到任何声音，可能需要安装 ALSA 工具。CAGE Web Design 的一个资料页面介绍了它的安装。

C.9　让树莓派说话

声音在树莓派上正常工作后，让它说话就不难了。首先，需要安装 pyttsx，这是一个 Python 文本转语音的库⑤。像下面这样安装它：

```
$ wget https://pypi.python.org/packages/source/p/pyttsx/pyttsx-1.1.tar.gz
$ gunzip pyttsx-1.1.tar.gz
$ tar -xf pyttsx-1.1.tar
$ cd pyttsx-1.1/
$ sudo python setup.py install
```

---

① "Working with SSH Key Passphrases"，GitHub 帮助，https://help.github.com/articles/working-with-ssh-key-passphrases/。
② 请参阅树莓派相机模块的产品页面，网址为 http://www.raspberrypi.org/product/camera-module/。
③ The RaspberryPiGuy，"Raspberry Pi-Camera Tutorial，"YouTube（2013 年 5 月 26 日），https://www.youtube.com/watch?v=T8T6S5eFpqE。
④ "Raspberry Pi—Getting Audio Working"，CAGE Web Design（2013 年 2 月 9 日），http://cagewebdev.com/index.php/raspberry-pi-getting-audio-working/。
⑤ 你可以在 https://github.com/parente/pyttsx/找到 pyttsx 的 GitHub 库，这是一个 Python 文本转语音的库。

然后需要安装 espeak，如下所示：

```
$ sudo apt-get install espeak
```

现在将扬声器连接到树莓派的音频插孔，并运行以下代码：

```
import sy:
import pyttsx

# main() function
def main():
    # use sys.argv if needed
    print 'running speech-test.py...'
    engine = pyttsx.init()
    str = "I speak. Therefore. I am. "
    if len(sys.argv) > 1:
        str = sys.argv[1]
    engine.say(str)
    engine.runAndWait()

# call main
if __name__ == '__main__':
    main()
```

C.10　让 HDMI 工作

你可以用 HDMI 电缆将树莓派连上显示器或电视机。为了确保它在第一次启动树莓派时工作，请在计算机中打开树莓派的 SD 卡，并在顶层目录中编辑 config.txt，如下所示：

```
hdmi_force_hotplug=1
```

现在，如果树莓派启动，你应该能够通过 HDMI 看到输出。

C.11　让树莓派移动

总是可以让树莓派使用电源适配器。但有些时候，你可能希望让树莓派移动，而没有麻烦的电线。为此，你需要一个电池组。一个效果很好的选项是带有兼容 micro USB 输出的可充电电池组。我用 Anker Astro Mini 3000mAh 外部电池取得了很好的效果，这可以在网上找到，约 20 美元。

C.12　检查树莓派的硬件版本

树莓派有几种版本。你可以登录到树莓派并在终端中输入以下命令，检查树莓派的硬件版本：

```
$ cat /proc/cpuinfo
```

下面是我的终端上的输出：

```
processor : 0
model name : ARMv6-compatible processor rev 7 (v6l)
BogoMIPS : 2.00
Features : swp half thumb fastmult vfp edsp java tls
CPU implementer : 0x41
CPU architecture: 7
CPU variant : 0x0
CPU part : 0xb76
CPU revision : 7

Hardware : BCM2708
Revision : 000f
Serial : 00000000364a6f1c
```

要了解修订版本号（revision），请参阅 http://elinux.org/RPi_HardwareHistory 上的硬件修订历史记录表。在我的例子中，是 2012 年第 4 季度制作的 PCB 版本 2.0 的 B 型。